T0179932

THE AMATEUR BIOLOGIST

SCIENTIFIC AMERICAN

THE AMATEUR BIOLOGIST

Edited by Shawn Carlson

JOHN WILEY & SONS, INC.

Published by John Wiley & Sons, Inc., New York
Published simultaneously in Canada

Figures 1.1, 4.1, 4.2 copyright © 1998 by Ian Worpole
Figures 10.4 copyright © 1996 and 19.1 and 19.2 copyright © 1995 by Michael Goodman
Illustrations on page 150 copyright © 1997 and Figure 21.1 copyright © 1996 by Bryan Christie

Library of Congress Cataloging-in-Publication Data is available from the publisher.

ISBN 978-1-63026-181-8

Printed in the United States of America

10 9 8 7 6 5 4 3 2 1

CONTENTS

FOREWORD

The Amateur Scientist

The book in hand, titled *The Amateur Biologist,* is the second of a series of books that derive from the longest running column in *Scientific American,* for the past 50 or so years known as "The Amateur Scientist." The origin of "The Amateur Scientist" goes back to a column first published in May 1928 and written by Albert G. Ingalls. It was called then "The Back Yard Astronomer." Ingalls' first sentence in that new column declared: "Here we amateur telescope makers are, more than 3000 of us, gathered together in our own back yard at last." At the top of the page is an illustration of the "Back Yard," with an amateur astronomer at work, apparently drawn by one Russell W. Porter, then considered a mentor of telescope makers throughout the land. Porter declared that the name of the new column, contracted to "Backyard Astronomer," conferred an honorary B.A. degree on all its readers.

It is fitting that this book series launched in early 2001, because the column in *Scientific American* has in this same year recently been discontinued to make room for a plethora of new features in the magazine. These volumes will enable the reader to continue to access many of the gems published in the column over the 70 years of its overall existence. The first volume in the present series was titled *The Amateur Astronomer* and was edited by Shawn Carlson, who also edited "The Amateur Scientist" column from 1995 to early 2001. Those of us at *Scientific American* are enormously grateful for Dr. Carlson's devotion to the column in its final period and his willingness to edit this present series, which we hope will preserve for the future many of the highlights of "The Amateur Scientist."

Long-time readers of *Scientific American* invariably recall favorite columns from "The Amateur Scientist." For example, as a former naval aviator one of my favorite pieces is from the April 1998 issue, on a home-built high-altitude vacuum chamber, to which I have referred a number of retired navy friends. As I recall the expensive equipment in U.S. Navy training centers, it seems to me amazing that a similar set-up can be replicated in one's garage. This particular column, originally titled "Making Experiments Out of Thin Air," seems to me an extremely informative piece, as are so many others that show up in this book and its series. (The column appears here as Chapter 4, "High Altitude Chamber," beginning on page 26.)

I am often asked about the readership of *Scientific American*—who these readers are and where their interests lie. My strong suspicion is that most of these inquiries come from potential readers who for one reason or another have not yet made the leap. Summarizing our readership in a brief reply is very difficult, because the population is very large and very diverse and includes isolated clusters of people who would not normally be identified as amateur scientists. If pressed for a brief reply, however, I often mention that my image of a reader of *Scientific American* is someone whose favorite fantasy on a rainy Saturday afternoon is to read "The Amateur Scientist" and to plan a future scientific experiment based on the information in the column. My sincere hope is that this series of books serves to fill the same precious moments.

John J. Hanley
Chairman Emeritus
Scientific American, Inc.

INTRODUCTION

I do love physics, but I wasn't a particularly good graduate student. In fact, I failed my first written qualifying exam badly. A second failure would have meant automatic expulsion from the University of California at Los Angeles and the end of my lifelong dream of becoming a physicist. So I spent the entire spring and summer of 1983 pounding the books fourteen hours a day. I worked every problem I could find, struggling to get extremely good at something physicists never actually do: solve problems without access to texts, under extreme emotional duress and in a limited time. I hated every minute of it.

Worse, each glance out my window made my torment more difficult to bear. For out there beckoned the whole natural world. Butterflies taunted me, dancing about the lilacs below my window as though they knew I was too busy to net and admire them. Neighborhood birds woke me up at daybreak with their impromptu symphonies, just as they always had. Earlier, I would often follow them into the fields and marvel at their territorial displays. But the chains to my books made these outings impossible and so this sweet music jolted me awake each day on a crashing wave of despair. And so it was for the blooming flowers that called out to be pressed and cataloged, and for the neighborhood pond paramecium ripe for a quick dance under my microscope, and for the fossils, unearthed by gentle summer rains, that were out there waiting to be found. The professional within me knew I was doing the right thing, but my inner amateur howled. I was miserable.

Then just five days before the test, a dear friend gave me something that he hoped would improve my mood; a wonderful book called *The Amateur Naturalist* by Gerald and Lee Durrell. It did improve my mood, and it almost ended (or perhaps it saved) my physics career. It was a big and beautiful book, full of lavish photos and meticulous drawings and over-brimming with how-to secrets. Every page made the pleasures of homebrew biology seem so appealing. I couldn't take it any longer. I decided to abandon physics on the spot and become a naturalist. I rushed back into the field and spent three glorious days in the national forests bordering Los Angeles, pecking about with my binoculars, my field notebook, a portable microscope, and, of course, the Durrells' wonderful book. I was absolutely thrilled with my new choice of vocation.

All this took the pressure off me. What a relief! I didn't think about the test again until the evening before it was scheduled to start. In fact, I only took the exam out of curiosity (or so I told myself at the time) just to see how I'd do. When I learned that I had passed, and my position at the university was secure, my sudden career move did seem a tad impulsive. Eventually I completed my doctorate and became a professional physicist. But my heart still sings whenever I indulge my amateur's passion for biology (and now it sings quite often).

So this volume, the second in a series of compendiums from the pages of *Scientific American*'s "The Amateur Scientist" column, has been a special pleasure for me to compile and edit. In fact, in the five-and-a-half years that I was privileged to write the column, I contributed many more projects from biology than physics. (*Scientific American* ended the feature's 73-year run in April 2001, when new management shifted the editorial focus of the magazine and suspended all of its long-running columns, including "The Amateur Scientist.") However, I've only included a few of my columns here so as to make room for a smattering of the many wonderful biology projects described by my predecessors at "The Amateur Scientist"'s helm: Albert Ingalls, Jearl Walker, and most especially C. L. Stong.

I've struggled to make this work the most comprehensive and wide-ranging collection of projects for the amateur biologist ever published in book form. This book contains projects to challenge the novice and experienced amateur alike; enough to keep any homebrew naturalist happy for a lifetime. It encompasses a wide array of fields including botany, genetics, behavior studies, cellular biology, microscopy, microbiology, and entomology. You'll even learn how to extract and purify DNA in your kitchen using common household materials. And to help you equip your home biological laboratory, I've included an extensive section that describes how to perform many of the core techniques of biology, as well as how to build important equipment that every biologist, professional or amateur, needs. But while this book has an impressive range of projects, any serious amateur will need a great deal more basic information about biology than would fit in one volume. You'll find some great additional sources of information in the "Further Reading" section.

This book digs deep into the archives and so some of these articles are three, even four, decades old. To make them more current I've interspersed quite a few comments about the text. You'll also find the supplier list to be radically updated. Sadly, these days few laboratory supply houses will sell anything to equip a home laboratory. That's partly why I founded the Society for Amateur Scientists (*www.sas.org*), to make sure that amateurs have access to the necessary materials of science. So if you can't secure a particular item elsewhere, the Society will do its best to get you what you need.

Although any responsible high-school student can safely carry out most of these experiments, it is essential that parents look over their youngsters' shoulders to make sure their children exercise reasonable care. You'll find warnings in the text that detail the risks and also spell out how to avoid them. Take a tip from the professionals and always make sure to follow the procedures detailed in these chapters that will keep you and your family safe.

This book wouldn't have been possible without the hard work of many people. Once again I must thank Diane McGarvey of *Scientific American*, who was the driving force behind this, and every compendium in this series. Her diplomatic aplomb kept this project from falling apart more than once. I'd also like to thank my wife, Michelle Tetreault, for her saintly patience with me throughout the laborious task of pulling this together and for her help with the glossary. I'd like to thank my two beautiful toddlers, Katherine and Erik, for providing me with hours of delightful distractions playing airplane and kissy games. They make me want to be a better human being. And I'm delighted to send a special thank you to those of my professors who chastised me for having "too much personality" and who sternly warned me that my outside interests proved that I wasn't really "serious" about physics. It would be the ruination of my career, one insisted. You guys gave me more encouragement than you will ever know! And lastly, I want to thank the Durrells for writing their transcendental work that saved my sanity and possibly salvaged my career. I can only hope that the present compendium will touch a single person as deeply as their book affected me. If so, then my labors will have been well worth it.

PART 1

MISCELLANEOUS TECHNIQUES

1 VIDEO MICROSCOPE

by Shawn Carlson, October 1998

A few years ago I decided to explore the microscopic menagerie living in a bit of rainwater that had collected in an open barrel. It proved to be a rich find. As the infinitesimal neighborhood of a single droplet came into focus under my microscope, I discovered many organisms I had never seen before. One in particular intrigued me. At first it looked like a cylindrical creature with a moving gut. But after a few moments I realized that I was peering at a tiny, tubular home being constructed by an even more tiny and highly industrious architect. The little worker trundled back and forth along the tube, snatching floating bits of organic refuse, which it then used to extend its domicile. I watched transfixed, all the while wishing I had some way to document this activity. Still photographs would have been wholly inadequate; the situation clearly required a video recording. But I didn't have a way to attach my camcorder to my microscope and thus was unable to share the odd antics of this aquatic charmer with others.

So you can understand why I was thrilled to receive a package from Charles Carter, a talented amateur scientist in London, Ontario, addressing just this problem. One can, of course, buy a commercial video camera and adapter specially built for this task, but these units are expensive. Besides, using a home camcorder has certain advantages. For example, it makes it easy for you to include a running commentary about your procedures and observations. And with the software and hardware now widely available, you can easily capture individual video frames on your computer for additional analyses.

Carter's invention consists of two parts: an adapter that optically links the camcorder with the microscope and a stand that holds the camcorder in place. Both pieces can be made in an afternoon for very little money.

If you lack a microscope, worry not. Microscopes often turn up where secondhand items are sold. For instance, you might scour your local thrift stores and pawnshops for a bargain. Carter found a simple monocular instrument at a garage sale for $10. I bought a research-quality binocular microscope from a friend for $100. And some brand-new microscopes are well within a typical amateur's budget. Small instruments can be purchased at many shopping malls for less than I paid. You can also consult the forum hosted by the Society for Amateur Scientists on the World Wide Web (*www.sas.org*) and check the online swap meet there for bargain equipment.

Carter's adapter couldn't be simpler. It uses a hood that screws around the lens of the camcorder to shade it from glare. Many camcorders just recess the lens in the housing for shading, but they still have threads for filters in front. So if your camcorder did not come with a separate lens hood, or if you don't want to sacrifice it for this project, you can probably find a lens hood of the proper diameter by rummaging through the junk box of your local camera shop. Alternatively, you can always buy a new one for a few dollars.

The eyepiece for most microscopes consists of two lenses situated at either end of a short metal tube. The top lens, the one you hold your eye near, is normally blocked off except for a hole at the center about the size of your pupil. This opening is too small for your camcorder to see through. The bottom lens, however, is larger and virtually unobstructed. And your camcorder will focus just fine with only this one lens in the eyepiece.

Because eyepieces sometimes need to be cleaned, the top lens is normally designed to unscrew from the tube. The other lens of the eyepiece may be attached more permanently, usually recessed slightly from the bottom end of the tube. Unscrew the top lens and discard it. You will also need to remove the outer housing that holds the eyepiece in place by unscrewing it from the main body of the microscope. Turn the eyepiece upside down and insert it into the eyepiece holder so that the lens projects above the top of the holder by about ¼ inch (about half a centimeter). Use a few drops of Krazy Glue to hold it in place.

Set the camcorder lens hood thread side down onto a wide strip of masking tape that is positioned sticky side up on a flat surface. Then place the eyepiece holder neck down in the center of the hood, pressing it firmly against the tape. Now mix a batch of epoxy and pour it between the neck and the lens hood. Take care not to allow any epoxy to ooze onto the threads of the lens hood. After the epoxy sets, lift the assembly and remove any tape sticking to the lens hood or covering the eyepiece.

Although you could now just screw the adapter to the microscope and the video camera to the adapter, that top-heavy arrangement would be quite precarious. Moreover, some high-end microscopes have their eyepieces canted to the side for the comfort of the viewer, and these instruments would not be able to support the weight of a camcorder attached at an angle. But Carter devised a sturdy stand that helps to hold his camcorder in position yet lets it freely slide up and down as he focuses the microscope. The stand functions best when the eyepiece is vertical, but it should also work in situations where the eyepiece must remain at a slight angle.

Carter built the base of his stand from a piece of scrap ¾-inch plywood to which he attached an adjustable closet rod. First affix the end of the closet rod to the base. Then place your microscope on the base, screw on both the adapter and your camcorder, and adjust the microscope's focusing knob so that the camcorder is as low as it can go. Take care and make sure things don't topple over at this point *[see Figure 1.1]*.

To attach the camcorder to the rod, Carter cleverly exploits the threaded sleeve on the base of the unit (where a tripod would normally screw in). While holding the microscope-camera combo upright with one hand, mark a line on the outer tube of the closet rod about one inch below the tripod mount. Then cut the outer tube off at that point with a pipe cutter or hacksaw. Next, cut the inner tube of the closet rod so that when it is inserted all the way in, it sticks out about four inches. If you have a stage-focusing microscope, you can dispense entirely with the inner rod and attach the camera directly to the primary support.

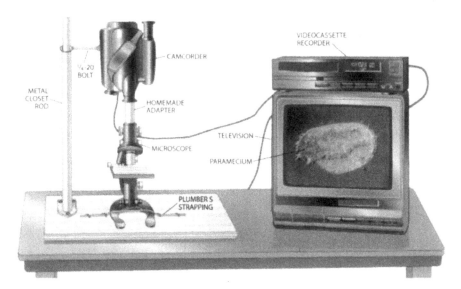

Figure 1.1   An inexpensive video microscope system, which allows several people to share the same view, can be built with a few dollars and an afternoon's work.

Next, drill a ¼-inch hole into the inner rod so that the camera can be secured using a long ¼-20 bolt. Slip a washer on the bolt, push the bolt through the rod, and add another washer and two nuts. Now screw the bolt into the camcorder and tighten the nuts, one against the closet rod and the other against the camera. Fix the microscope to the plywood stand by securing it with a length of plumber's strapping across the base. Now you're ready to spy on an invisible realm. To use your new apparatus, focus the microscope until a clear image appears on your monitor and then adjust the zoom on the camcorder until the image fills the screen.

Although this system will enable you to take leisurely safaris through microscopic jungles, it may be difficult to gauge the size of your minute prey. The best way to address this problem is to measure on your monitor the dimensions of something of known size. I invite professional and amateur microscopists to share other suggestions for calibrating this instrument with the Society for Amateur Scientists by joining the discussion on the society's web site.

2 ELECTROPHORESIS

by Shawn Carlson, December 1998

The most wonderful private garden I have ever seen is tucked away behind a modest house in La Jolla, California, not far from where I live. The gardener is a British-born psychology professor and dear friend who sends me home with fruit and flowers each time I visit. Recently I noticed that two of his plants, though very different in shape, produced flowers of the exact same shade of purple. This observation made me wonder whether the two species might be related.

One normally traces evolutionary connections by identifying physical similarities between species. So I decided to extract and isolate the pigments in the two flowers so that I could compare them in detail. That process is actually much easier than it sounds. In fact, using a simple technique called electrophoresis, I could carry out the experiment in about an hour for very little money.

Most molecules are electrically neutral, but some important biological molecules, including proteins, DNA fragments, and many natural dyes, carry a net negative charge when they are in solution. Electrophoresis cleverly uses a weak electric field to force such charged molecules to drift through a medium that separates them by offering differing amounts of resistance to motion.

You can easily see this phenomenon in action when you place a droplet of dye on a strip of blotting paper that has been wet by a conductive fluid, such as salt water. When the ends of the paper are connected across a battery, a voltage is set up, which drives the charged dye molecules through the paper. Positively charged particles move toward the negative terminal, whereas negative ones move toward the positive terminal. Usually, larger molecules have a more difficult time than smaller ones in passing through paper fibers, so the smaller molecules drift faster. Thus, over time, the different molecules in a mixture will tend to sort themselves by size.

It takes only a few minutes to set up a basic apparatus. From a large coffee filter cut a rectangular strip of paper that is about 1 centimeter (about half an inch) wide and about 15 centimeters (6 inches) long. Place this paper band inside a flat glass pan or cooking dish. Roll each end of the paper strip around a nail, and use an alligator clip to secure it. Wire the clips to five nine-volt batteries connected in series.

To make the conductive solution, mix about 100 milliliters (4 ounces) of distilled or bottled water with 1.5 grams (about ¼ teaspoon) of table salt. Then thoroughly wet the paper, including the nails, with the salt solution, but don't add so much that the paper is submerged in a puddle.

To begin, use a toothpick to place droplets from several different hues of food coloring in a line, then connect the electrodes. The colors will rapidly spread into streaks as the pigment molecules migrate toward the positive electrode. Next, mix two of the dyes, say, red and green, and run a tiny splotch of the combination. After about 20 minutes, the colors should begin to separate. The same technique can be used to separate other molecular mixtures.

So here's how to find out if two plant species use the same molecules as pigments. First, crush the flowers and immerse them in clear isopropyl alcohol, letting the solids settle. Pour off each of the resulting color-tinged liquids into separate containers and then concentrate them by letting the alcohol evaporate. Once the alcohol is nearly gone, dissolve the pigments in a few drops of the salt solution you made earlier.

Next, line up three tiny dots of pigment on a strip of soaked filter paper by placing a pure sample from each plant on the outside and an equal mixture from both in the center. Then connect the batteries. If the outside dots separate into different sets of colored swaths and the center streak appears to be a combination of the outer ones, then you know that different pigments are involved. But if all three dots form the same pattern, then both plants probably rely on the same molecules for color.

Note that the salt ions will also drift toward the electrodes, where they will quickly create a layer of tarnish that impedes the flow of electricity. So after each run, you will have to scrub the electrodes. As all this cleaning rapidly becomes tiresome, you might try to replace the steel nails with another conductive material that does not tarnish as quickly—stainless-steel wire or aluminum foil, for example. Small pieces of gold or platinum wire or chain work especially well.

Although many great discoveries have been made using paper-based electrophoresis, this simple method does have a big drawback: the molecules tend to get caught up in the fibers of the paper. This complication explains why even pure dyes form streaks instead of remaining well-defined dots as they move along. So these days biologists often replace the paper with a more uniform material called agarose—a clear substance with the consistency of stiff gelatin. The DNA "fingerprint" patterns you may have seen are produced by electrophoresis on such a gel. Each of the individual lines in the fingerprint indicates strands of DNA of a certain length. Compared with results with paper, the degree of separation possible with a gel electrophoresis is amazing.

An earlier "Amateur Scientist" column explained how to extract DNA from living tissues. *[See Chapter 25 of this book.]* Unfortunately, the extracted material must be subjected to sophisticated laboratory manipulations using expensive reagents before a fingerprint can be created. But similarly diagnostic patterns can be made using plant pigments. Indeed, complex pigments often separate so cleanly that the results are just as stunning. After about 20 minutes, you can often isolate virtually every molecule involved in such a mixture.

Although ordinary gelatin does not work well, I'm told that a food additive called agar-agar may and that it can be found in Chinese food markets. But I suggest that you spend $25 and purchase enough agarose gel for about 40 experiments from Edvotek, an educational biotechnology company in West Bethesda, Maryland (301-251-5990 or 800-338-6835); *www.edvotek.com.*

You can quickly fashion a gel-based electrophoresis unit from any small, rectangular container that is waterproof. I used the bottom of a plastic soap dish. Bend some aluminum foil over the two shorter sides to serve as electrodes.

The secret to successful electrophoresis is in the buffer solution. If it is too conductive it will carry too much current and heat and distort the gel. Walt Allen of the Foundation for Blood Research in Scarborough, Maine, submitted the perfect recipe for the buffer: mix 1.5 grams (¼ teaspoon) of baking soda to 250 milliliters (10 ounces) of tap water.

Then pour enough of the hot, liquid agarose into the dish to cover it with a half-centimeter layer. Because your gel must contain reservoirs to hold the concoctions you wish to separate, cut out a comb shape *[see Figure 2.1 on page 10]* from a Styrofoam tray—the kind used to pack meat at the grocery store—and suspend it so that the tines penetrate the liquid agarose but don't poke through the bottom. Let the gel set before carefully removing the comb. This maneuver should produce a series of nicely spaced wells for your samples.

For the separation to take place, the gel must contain ions that can

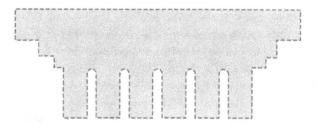

Figure 2.1   This template allows the "comb" to be cut from a Styrofoam meat tray.

conduct electricity. To add the ions, make the gel with the buffer. Allen's recipe mixes 1.5 grams (¼ teaspoon) of the agarose gel into 50 milliliters (2 ounces) of buffer.

Next, the pigments need to be dissolved into what's called a loading solution before they are placed in the wells. Allen dries the pigments, then mixes them with a solution consisting of three parts glycerol and seven parts buffer *[see Figure 2.2]*.

In addition, Allen makes the following recommendations: Don't let the gel occupy the entire container. Rather, pour the gel with pieces of Styrofoam fit in the ends of the soap dish. Allen uses Styrofoam from a 1-centimeter-thick sheet of packing material to fit in the ends. Once the gel has set and the comb has been removed these are removed so that the gel sits in the middle of the container.

Second, cover the gel with buffer to a depth of 3–4 mm filling the buffer reservoirs at each end of the gel (where the Styrofoam was removed). This keeps the gel from being distorted by the hydrolysis that takes place at the aluminum foil electrodes.

And lastly, use an eyedropper to drop the sample into the wells. You do this by just penetrating the buffer surface and allowing the glycerol-thickened sample to fall into the wells. When using a pigmented sample this is not difficult to do since you can see it drop and any that does not fall in the well will diffuse away in the buffer. Make your wells small (about 2 mm thick) to keep the sample volume down and concentrated at the bottom of the wells. Rinse the dropper thoroughly between samples.

To start your experiment, just connect the aluminum foil to your batteries with alligator clips, with the positive terminal attached to the side opposite the wells so that the negatively charged molecules have some room to move. Don't worry if you notice some bubbling along the foil as water molecules are split apart by electrolysis. And don't be concerned if the color of the pigments changes (a common effect of altered pH).

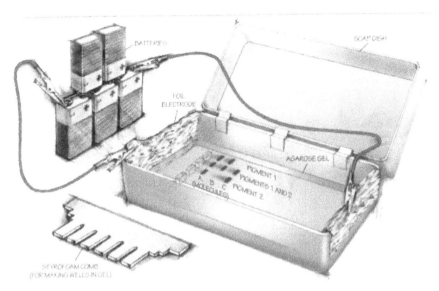

Figure 2.2   A soap dish with a layer of agarose gel permits complex molecular mixtures to be separated by electrophoresis.

Because of its tendency to tarnish, you will have to replace the aluminum foil when you renew the agarose after each run.

Electrophoresis is a cornerstone of molecular biology. Armed with this technique you can isolate the basic stuff of biology for further exploration. There are far too many living systems for professionals to study them all, and so there are many discoveries waiting for the ambitious amateur armed with this technique, a textbook, and some perseverance. So why not get to work!

The Society for Amateur Scientists has joined forces with Edvotek to create a complete gel-based electrophoresis unit for kitchens and classroom labs. Send $55 to SAS, 5600 Post Road, #114-341, East Greenwich, RI 02818, or call the society at 401-823-7800. You will find more information about this and other articles from the Society for Amateur Scientists on the World Wide Web (*www.sas.org*).

3 MEASURING METABOLISM OF SMALL ANIMALS

by C. L. Stong, July 1969 and August 1957

A person lives far longer than a mouse, but in one sense they come out about even, as can be verified with a homemade instrument for measuring metabolism. The average human weighs more than 500 times as much as the average mouse, and of course people need more food, but pound for pound they process food and expend energy at about the same rate. Jean K. Lauber, who is assistant professor of zoology at the University of Alberta, refers to metabolic rate as a measure of the "aliveness" of animals. She has designed an easily constructed apparatus for measuring metabolic rate and suggests a series of experiments that can disclose some interesting facts about the chemistry of animals.

"Metabolism," she writes, "is essentially an oxidation process. The animal uses oxygen in direct proportion to the rate at which it burns food-stuffs for releasing energy to run its internal machinery—energy for growth, for keeping the organism warm, for movement, and for running the myriad chemical reactions that constitute the life process. For this reason the amount of oxygen consumed by an organism per unit of time can be used as a measure of its metabolic rate.

"Oxygen consumed by an organism can be monitored in a number of ways. Most hospitals, for example, determine metabolism with a spirometer, an instrument that responds to the volume of oxygen consumed by the patient. Exhaled carbon dioxide, the principal waste product of the metabolic process, is removed chemically. The apparatus includes a pen recorder that automatically draws a graph of the rate at which oxygen is supplied to the patient through a facemask.

"A number of simpler devices have been made for determining the metabolism of small animals. Most of the instruments embody the same basic principles. Oxygen is measured in terms of volume consumed per unit of time. Carbon dioxide is removed, sometimes being measured. Certain variables must be taken into account. For example, the animal must be fed normally so that the results of the experiment are not influenced by a diet that is deficient or excessive.

"The rate at which oxygen is consumed reflects the physiological state of the whole animal, the sum total of its chemical reactions. For this reason the experiment can disclose the influence of drugs on a selected chemical reaction within the metabolic scheme. The more significant variables are temperature and barometric pressure. The entire experiment must be run under conditions of controlled temperature and pressure, or these variables must be taken into account mathematically when the data are reduced. The environmental conditions may change substantially during an experiment. For example, marked changes in both temperature and barometric pressure may occur with the approach of a storm. Variables of this kind can be measured by including a thermobarometer in the experiment. It is an apparatus exactly like the one that measures metabolism, but it contains no organisms. Any changes of volume that occur in the thermobarometer system are entered as corrections in the data from parallel runs in which organisms are used.

"One of the most accurate and widely used devices for monitoring oxygen consumption was devised by the German physiologist Otto Warburg. As I have modified it for amateur construction it consists of a sealed animal chamber fitted with a slender glass tube that is open at the outer end *[see Figure 3.1]*. As oxygen is consumed by the animal more air flows into

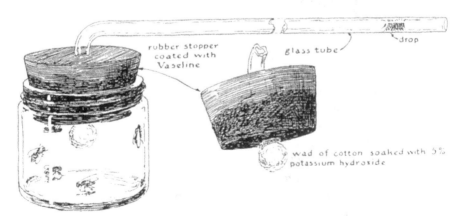

Figure 3.1 Elements of an apparatus for measuring oxygen consumption.

the chamber through the manometer tube, carrying with it a droplet of colored water or a strong film of bubble solution to serve as a visual indicator of the gas flow. Carbon dioxide exhaled by the animal is absorbed by a solution of potassium hydroxide on a small wad of cotton. I attach the cotton to the inner end of the tube with a rubber band and then moisten the wad with solution by means of a medicine dropper. *[For a Warburg apparatus adapted to monitor the metabolism of insects, see page 153. Ed.]*

"The potassium hydroxide, which represents 5 percent (by weight) of the solution, is a form of lye that is both toxic and corrosive. Handle it accordingly. If the solution comes in contact with the skin, wash the affected part immediately with lots of water and rinse with vinegar. The chemical will also burn animals on contact. If the animals can crawl up the sides of the bottle or can fly, I enclose the chemical in a protective cage of wire screening.

"I have used this simple apparatus to measure the oxygen consumption of a large variety of small animals, including houseflies, honeybees, fruit flies, earthworms, grasshoppers, crayfish, clams, and even fish. Almost any animal can be studied if it is small enough to fit into a wide-mouthed bottle that has a capacity of one or two ounces. Usually I do several experiments simultaneously, using a separate chamber for each animal or group of animals.

"Each bottle is closed with a rubber stopper that is perforated to make a snug fit with a glass tube 1 to 3 millimeters in inside diameter. The manometer tube should be about 50 centimeters long, bent to a right angle 2 centimeters from the stopper so that the long segment extends horizontally when the bottle stands upright. The side of the stopper is coated lightly with Vaseline. After one or more animals have been put in each bottle the stoppers are inserted and held in place with a wrapping of waterproof adhesive tape. The bottles are placed in a wire rack that is immersed in a water bath so that the necks of the bottles are just at the waterline. The water has been previously heated or cooled to the desired temperature. If necessary, support the outer ends of the tubes so that they are approximately horizontal.

"After 15 minutes, when the temperature of the air in the bottles has reached the temperature of the surrounding water, I start the experiment by placing a drop of indicator solution (water tinted with food coloring) in the open end of each tube.

"As the animal continues to respire, oxygen is drawn into the bottle. The volume represented by each one-centimeter length of the tube, multiplied by the distance traveled by the indicator drop as the experiment proceeds, gives the volume of oxygen used. To find the volume of the tube per centimeter of its length, measure the inner diameter of the tube as accurately as possible in centimeters, divide by two and multiply the quotient by itself and by 3.1416. Expressed algebraically, the formula for the volume is $V = \pi r^2 h$, in which V is the volume per centimeter of tube length, π is 3.1416, r is the radius of the tube, and h is its height (in this case one centimeter).

"Include in the rack of immersed animal chambers one empty chamber of identical construction to serve as the thermobarometer. Place the indicator drop inside the thermobarometer tube a few centimeters from the outer end of the tube so that it can move in either direction. Mark the position of the indicator on the glass with a grease pencil. With the apparatus and animals thus prepared, record on a sheet of paper the time when the indicators are placed in the manometer tubes; on the same line enter a zero at the head of a separate column for each bottle in the rack, including the thermobarometer. At 10-minute intervals for one-hour record for each tube the distance in centimeters that the indicator has moved from its initial position.

"Measure and record the temperature of the water bath in degrees Kelvin. To get degrees Kelvin add 273 to degrees Celsius. If your thermometer is calibrated in degrees Fahrenheit, subtract 32 from the reading, divide the remainder by 9, multiply the quotient by 5, and add 273 to get degrees Kelvin.

"Measure and record the barometric pressure in torr. If your barometer is calibrated in millimeters of mercury, simply record the reading. (One torr is equal to the pressure exerted by a column of mercury 1 millimeter in height.) If the instrument is calibrated in inches of mercury, multiply the reading by 25.4 to get the pressure in torr.

"From these data you can find by a few simple calculations the amount of oxygen each animal consumes. The actual consumption may vary with the pressure of the atmosphere and with the temperature. It is useful for comparing experimental results observed at various times and under various conditions to adjust the data to a standard barometric pressure and temperature. By agreement biologists use as standards a barometric pressure of 760 torr and a temperature of 273 degrees K.

"First find the apparent volume of oxygen consumed by the animal or animals in each bottle during each of the six 10-minute intervals. To find the apparent oxygen consumption in cubic centimeters multiply the distance in centimeters that the indicator moved during each 10-minute interval by the previously determined volume per centimeter of the length of the tube. Determine for each interval of time the change in volume of the air that took place in the thermobarometer. The volume of air in the thermobarometer may have increased (indicated by the outward movement of the colored drop or the bubble film), decreased (indicated by the movement of the indicator toward the bottle), or both.

"Correct the apparent oxygen consumption to the actual consumption by adding or subtracting from each figure the change that occurred simultaneously in the volume of air in the thermobarometer. Record the actual consumption of oxygen for each bottle during each of the six time intervals. Finally, adjust the actual consumption to standard pressure and temperature.

"For each bottle and time interval multiply the actual consumption of oxygen in cubic centimeters by .36 and by the barometric pressure in torr and divide the product by the recorded temperature in degrees Kelvin. Expressed algebraically, the conversion formula is $V' = .36VP/T$, in which V' is the volume of oxygen consumed at standard pressure and temperature, V is the actual volume as measured by the experiment, P is the barometric pressure indicated by the barometer and T is the observed temperature of the water bath in degrees Kelvin. Record the oxygen consumption as corrected for standard pressure and temperature.

"As the final step weigh the animal in grams. Add the corrected volumes of oxygen consumed during the six intervals to get the total consumption for one hour and divide the total consumption in cubic centimeters by the weight of the animal in grams to determine the consumption of oxygen per hour per gram of the animal's weight at standard pressure and temperature. A graph can be drawn to show the oxygen consumption of each experimental animal during the experiment by plotting the oxygen consumption, in cubic centimeters, against time in minutes.

"Many variations of the above experiment suggest themselves. One can compare the performance of a single large animal with that of several smaller animals of the same total weight. Similarly, the effects of increased or reduced amounts of light can be measured, but in this experiment take care to maintain the water bath at constant temperature. One animal of a pair can be restrained in a cheesecloth bag while its companion remains free in a separate bottle, thus demonstrating the effects of reduced activity on metabolism.

"Most invertebrates exhibit a decrease in metabolic rate with reduced temperature. To lower the animal's temperature immerse one bottle in an ice bath. If aquatic animals are used, each animal chamber and the thermobarometer should contain a standard amount of the water in which the animal normally lives. The chambers should not be more than half filled with water, however, because the air above the water must contain enough oxygen to supply the animal normally during the experimental run.

"For animals as large as a mouse the equipment that has been described works nicely if the experiment is limited to about 10 minutes. The animal chamber and the manometer tube must be of appropriate size or the mouse will use up all the oxygen and suffocate. Watch the movement of the soap film carefully during the experiment. If the movement of the film slows appreciably, the animal may be running out of oxygen. In this event remove the bottle from the water bath and open it immediately. Then substitute a manometer tube of about twice the diameter of the first one and try again.

"It is also easy to construct a scaled-down version of the spirometer, the instrument used for determining the metabolic rate of humans and other large mammals. The scaled-down instrument *[see Figure 3.2 on page 18]* consists essentially of an airtight animal chamber connected by tubing to a reservoir that is in turn linked mechanically to the stylus of a kymograph, which automatically draws a graph displaying oxygen consumption against time. *[You'll find how to construct a kymograph from a tin can in Chapter 5. Ed.]* Apparatuses of this type can be designed for accommodating animals weighing from 50 to 1,000 grams simply by varying the size of the animal chamber and the oxygen reservoir.

"For approximating the dimensions of the oxygen reservoir a good rule of thumb is to assume that each hour an animal at rest will require about one cubic centimeter of oxygen per gram of body weight. For example, a spirometer that contains 250 cubic centimeters of oxygen would be adequate for a one-hour test of a 250-gram animal. The accuracy of the measurements will suffer if the apparatus is made excessively large, because the animal will then consume only a small fraction of the available oxygen, thus reducing the change in volume of the spirometer and the resulting excursion of the stylus.

"The apparatus has been used for rats, mice, and baby chicks. The animal chamber is a wide-mouthed, one-pint Mason jar fitted with a self-sealing lid and screw band. Two ¼-inch holes were drilled in the metal lid, and a 1-inch length of ⅜-inch copper tubing was soldered in place over each hole. One copper nipple is connected by rubber tubing to the oxygen supply; the other one is the exhaust and is fitted with a short sleeve of rubber tubing so that this opening can be sealed by inserting a thermometer. It is extremely important that the system be leak-free. Grease the top of the jar and all rubber-tubing connections lightly with Vaseline, clamp on the oxygen-supply tube and check the apparatus for leaks by immersing it in a tub of water.

"Circles of ¼-inch-mesh hardware cloth were cut to fit the inside of the animal chamber. During the experiment they rest on top of a layer of soda

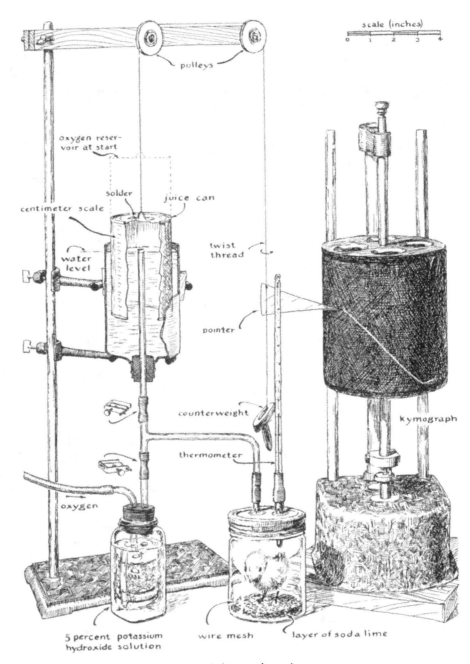

Figure 3.2 Jean K. Lauber's scaled-down spirometer.

lime ($NaOH + Ca(OH)_2$), which acts as the absorbent of the carbon dioxide. The stack of wire disks protects the animal from contact with the corrosive chemical; the mesh must be fine enough so that the animal's feet will not slip through the stack and touch the soda lime. For the oxygen supply I have used a tank or lecture bottle; if such a supply is not at hand, one can have a balloon or an inner tube, filled with oxygen at a local chemistry laboratory, hospital, or welding shop. [Any well-stocked hardware store will carry small oxygen bottles for home welding and brazing. The bottles cost under $10. Ed.]

"The oxygen supply is connected to the apparatus by flexible tubing and a T fitting, with the animal chamber on the lower arm and the spirometer on the upper arm. The spirometer is made from two telescoping tin cans, one slightly smaller than the other. The outer can of my apparatus is a liquid-soap container of the type that has a plastic neck and screw cap. The bottom of the can was cut out with a can opener. If one uses a different kind of container, one end is cut out and a 10-millimeter hole is drilled in the other end. This hole, or the neck of the soap can, is fitted with a one-hole rubber stopper. A glass tube extends through the stopper to a point 10 millimeters below the open end of the can. The can is mounted securely by a large clamp attached to a ring stand and is filled with water to approximately 20 millimeters from what is now the top. On the outside of the smaller can (I used a frozen-juice can, which clears the soap can by about 3 millimeters on all sides) a loop of wire is soldered in place exactly in the center of the bottom. The can is inverted inside the larger one and is suspended from the loop by a thread.

"The thread runs up over two pulleys. The first one is about 25 centimeters above the top of the outer can; the second pulley is at the same level but about 15 centimeters to one side. The thread runs around this pulley and down to a counterbalance equivalent in weight to the inner can. A centimeter scale marked on the side of the inner can is convenient. The thread on the counterbalance side carries a pointer that is cut from a piece of photographic film and is held in position between two knots in the thread. The pointer is brought in contact with the smoked surface of a kymograph drum, the thread having been twisted in such a way that its tendency to untwist will keep the pointer in constant but delicate contact with the drum.

"At the beginning of a run the inside can is moved to its lower position and a pinch clamp is placed between the spirometer and the upper arm of the T fitting. The chamber is charged with fresh soda lime and the animal is sealed inside. With the exhaust tube open, oxygen is first bubbled through water, a step that saturates the gas and also serves to indicate rate

of flow, and then is delivered to the chamber. After a five-minute equilibration period the clamp below the spirometer is removed and the spirometer is charged with oxygen by briefly closing the exhaust tube. The inner
can should now be in its upper position, but its lower rim must extend into
the water so that no room air can leak in to dilute the oxygen. The experimental run is started by simultaneously sealing the exhaust tube and
clamping the oxygen-supply tube.

"As the animal uses oxygen the inner can of the spirometer moves
down and the pointer writes a rising curve on the kymograph drum. Pressure in the animal chamber and the spirometer remains at atmospheric
level since the inner can of the spirometer is freely movable. If the animal
seems easily disturbed by activities in the room, drape a black cloth
around the chamber. On a day that is warm but not humid, cooling of the
chamber can be effected by dampening the cloth occasionally. It may also
be necessary to submerge the entire chamber in a constant-temperature
water bath; a sink or a plastic dishpan will serve for this purpose. Temperature fluctuations take place much more slowly in water than in air.

"It is necessary to know the volume of oxygen contained in the spirometer can. The volume can be determined by calculating the volume of a
cylinder, as in the previous experiment. An even easier way is to fill the
inner can with water and determine the depth in centimeters of a given
volume of water. For instance, in my setup 325 cubic centimeters of water
filled the can to the 10-centimeter mark. Thus each centimeter on the side
of the spirometer can, or each centimeter in the vertical rise of the recording stylus, represents a volume of 32.5 cubic centimeters. This is the
spirometer-calibration factor. It is also necessary to know how fast the
kymograph is turning, that is, how much time is represented by 1 centimeter of the base line of the graph. An easy way to get this figure is to
mark the exact starting point with an arrow on the kymograph tracing;
mark the finish point one hour (or fraction thereof) later.

"In a typical experiment I used weanling male rats divided into three
groups of two rats each. The animals always had access to food and water.
They were weighed daily, at which time the cages were cleaned. The control group, *A*, received a standard diet. To the diet of group *B* thyroid hormone was added at a rate of one crushed five-grain tablet of U.S.P. thyroid
per 40 grams of feed. Adequate mixing was assured by placing the diet plus
ground thyroid hormone in a large paper bag and shaking it well. Group *C*
rats received standard feed to which was added one crushed 50-milligram
tablet of propylthiouracil, an antithyroid drug, per 700 grams of feed.
(These drugs are prescription items, available only from a licensed phar-
macist. I have checked with several pharmacists in my area and have been

assured that an amateur biologist could probably obtain enough pills for such an experiment by explaining his need to his physician.)

"After one week on these diets weight differences became apparent. Differences in behavior were also beginning to appear: rats receiving the hyperthyroid diet *B* were definitely more active and 'jumpy,' whereas rats on diet *C* seemed sluggish. After two weeks on the experimental diets food was withheld overnight, and the following morning oxygen consumption was determined for each animal. Each rat was allowed a five-minute equilibration period in the animal chamber, and then oxygen consumption was monitored for 20 minutes. The chamber was washed, dried, and charged with fresh soda lime after each run.

"Barometric pressure on the day of the experiment was 701 torr. Other raw data were recorded in a table that showed in its first four columns the diet given the rat, the number of the rat, the rat's weight in grams, and the temperature of the animal chamber in degrees Celsius. There followed for each rat four columns. Column 1 gave the measurement from base line to finish point on the kymograph tracing. Column 2 was obtained by multiplying the figures in column 1 by 32.5, the spirometer-calibration factor. Column 3 was obtained by multiplying these figures in turn by 3 (20-minute run times 3 is 60 minutes). Column 4 was derived by dividing by the body weight. To reduce these figures to standard pressure and temperature (column 5) we used the same formula that was employed in the earlier experiment.

"The hyperthyroid diet *B* markedly affected metabolic rate; the loss of weight on this diet arises because the rats run off food reserves instead of storing them. The antithyroid diet *C* did not have such a dramatic effect on weight, but it depressed oxygen consumption. At the conclusion of the experiment all the rats were given a standard diet, and within a week they had returned to normal weight and behavior.

"In retrospect it would have been preferable to conduct all runs at exactly the same temperature. The conversion to standard pressure and temperature corrects for the expansion of oxygen in the spirometer that accompanies a rise in temperature, but it does not take account of another variable: all warm-blooded animals must expend energy, hence oxygen, in order to maintain a constant internal temperature in spite of varying external temperature. To this extent such determinations of basal metabolic rate are not quite basal, and the results should always be stated as cubic centimeters of oxygen consumed per gram per hour at a stated temperature."

A second amateur sojourn into animal metabolism was carried out by Nancy Rentschler, a high school student in Mayfield, Ohio. While looking for a science fair project she came across a textbook diagram of an apparatus to measure animal metabolism. She decided to adapt the design to measure the metabolism of mice.

She writes: "The mice I used were purchased through a pet shop. For the purpose of my experiments I divided 15 mice into four groups, three in one group and four in each of the others. By placing each group on a diet or medication which differed from that of the others, I could study the effects of these differences on the metabolism of the animals. I followed the experimental method devised by the noted British physiologist J. S. Haldane in 1890. The apparatus consists mainly of an animal chamber and five flasks of chemicals interconnected by tubing so that a controlled stream of air can flow through the system *[see Figure 3.3]*. The purpose of the apparatus is to measure the amount of oxygen taken up by the animal, and the amount of carbon dioxide expelled. The ratio of oxygen inhaled to carbon dioxide exhaled by the animal during a given period indicates the rate of its metabolism, and is called the 'respiratory quotient.' This quotient varies with the diet of the animal. When the animal is fed a carbohydrate such as sugar, the ratio is 1. When it is fed fats, the ratio varies slightly with the composition of the fat but averages 0.7. The ratio for proteins also varies, but averages 0.8. The ratio of alcohol is 0.667. The respiratory quotients of normal animals under average conditions usually lie between 0.72 and 0.97.

"Each flask of the apparatus is fitted with a rubber stopper and two glass tubes about half an inch in diameter. One tube reaches to within an inch of the bottom of the flask and the other just passes through the stopper. Air entering the flasks through the longer tubes is exhausted through the shorter ones. The first and fourth flasks in the series (not counting the animal chamber) are filled to a depth of about three inches with soda lime, which absorbs carbon dioxide. The second and third flasks contain the same amount of calcium chloride."

[At this point Stong's original article recommended creating a special surface to absorb water vapor expired by the animal by heating a pumice stone with an acetylene torch until it was red hot and dropping it into sulfuric acid. Oh boy! Fortunately, there are much safer ways to remove the water vapor. The most effective method is to pass the air through a "cold trap" made by submerging a U-shaped metal pipe into a bath of dry ice and alcohol. Look up "dry ice" in your Yellow Pages to find a source near you. Ed.]

"The soda lime is prepared by mixing lime with a solution of sodium hydroxide in the proportion of 1 ounce of sodium hydroxide (by weight) to 2½ ounces of water (by volume). Lime is added until the mixture becomes dry. The powder is then separated from the coarse particles by means of a fine sieve and discarded. Large lumps are broken down. It is the interme-

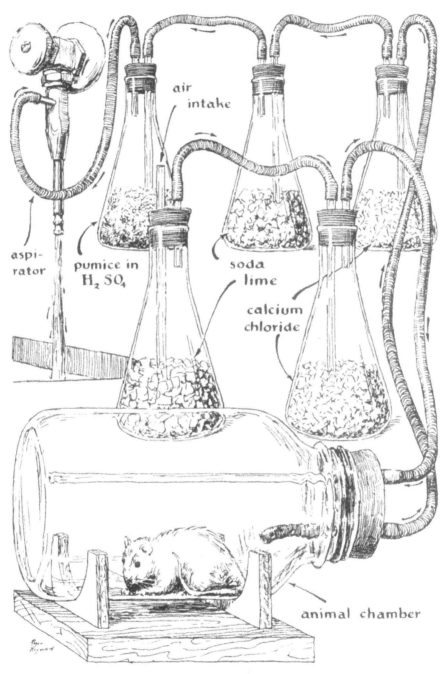

Figure 3.3 An amateur's apparatus for measuring the metabolism of mice.

air intake

aspi-
rator

pumice in
H₂ SO₄

soda
lime

calcium
chloride

animal chamber

diate fragments—those which pass through a sieve of five meshes per inch—that are used for charging the flasks. The absorbing power of soda lime does not last long, and I had to make additional batches as the experiments progressed.

"My animal chamber was a two-quart canning jar. I found it necessary to shield the exhaust tube of the chamber to keep it from pinning the mice. Before I added the shield, this happened several times, spoiling the experiment and injuring the mouse. The shield is merely a short length of rubber tubing with a slit or a few holes cut in it. It is slipped over the shorter glass tube inside the chamber. No damage is done when a mouse brushes against the end of the tube because the slit provides a second exhaust port.

"The entire system must be airtight. Close-fitting stoppers should be used and all joints coated with either wax or plastic cement. The rubber tubing should be as short and straight as possible, and should be tightly fitted to the glass tubes. Air was pulled through the apparatus by means of an aspirator attached to a water faucet.

"Air normally contains about 3 percent carbon dioxide and a varying amount of water vapor. Both are removed by the first and second flasks. Thus air free of water vapor and carbon dioxide flows into the animal chamber. The animal inhales oxygen and exhales carbon dioxide and water vapor. The latter are absorbed by the remaining flasks. The increase in weight of the third flask indicates the amount of water vapor given off by the animal. The fourth and fifth flasks measure the amount of carbon dioxide (which reacts with the soda lime in the fourth flask to form carbonic acid). The fourth and fifth flasks must be weighed together because the soda lime may give up moisture to the dry air and thus lose weight.

"In setting up the apparatus for a test run, the last three flasks are weighed, the fourth and fifth together. The animal is then placed in the chamber, which is stoppered and weighed. The test run is timed from this moment. The chamber is now connected to the apparatus and the air pump started. I ran the mice in each group for a total time of one hour. At the end of this period the pump is stopped and the chamber removed from the apparatus, stoppered, and weighed again. The third, fourth and fifth flasks are also weighed.

"The respiratory quotient may now be calculated. The combined weight of the mouse and chamber at the beginning of the run minus their weight at the end of the run equals how much weight the mouse has lost. The weight of the third flask at the end of the run minus its weight at the beginning equals the amount of water absorbed by the calcium chloride and lost by the mouse. The weight of the fourth and fifth flasks at the end of the run minus their weight at the beginning equals the amount of carbonic acid formed. The total weight of water and carbon dioxide absorbed minus the loss in weight of the mouse equals the weight of oxygen absorbed. The respiratory quotient is determined by multiplying the weight of the carbonic acid by the fraction $\frac{32}{44}$ and dividing the result by the weight of oxygen absorbed. The quantity $\frac{32}{44}$ is ratio of the molecular weight of oxygen to that of carbon dioxide. Its use in the equation indicates the amount of carbon dioxide represented by the carbonic acid.

"I used two of my four groups of mice to study the effects of diet on metabolism. With the other two groups I instigated the metabolic effect of the activity of the thyroid gland. The first group of four mice was given only water. Although mice normally live about nine days without food, these died after four days. It is likely that they contracted pneumonia because their resistance was low. Their respiratory quotient dropped slightly from the beginning of the experiment but stayed within the normal limit of .7 to 1 for the first three days. It plunged sharply just before the animals died. Oxygen consumption, however, decreased at a constant rate throughout the period of observation."

4 HIGH ALTITUDE CHAMBER

by Shawn Carlson, April 1998

Every creature on the earth lives under the warm, nurturing, and protective blanket formed by the atmosphere. Yet all this air does more than trap the sun's heat and carry gases between plants and animals. It also presses down on our world with powerful force. At sea level, a single sheet of writing paper, when laid flat, sustains 6,111 newtons—about 1,400 pounds.

One might imagine that such a burden would stress living creatures enormously. But far from hurting organisms, the weight of the atmosphere proves absolutely essential for life. Liquid water could not exist on the earth were not atmospheric pressure sufficient to keep it from boiling rapidly away. And many vital biological processes, cellular respiration chief among them, fail if the air pressure falls too low.

Of course, atmospheric pressure decreases with altitude, and the earth's surface pokes up quite high in many places. Remarkably, humans can adapt to almost any elevation on the planet. Few other species can thrive both along the coast and between the peaks of the Himalayas, nearly six kilometers (3.7 miles) above sea level. By exercising some smarts, humans can keep warm and fed even in cold, harsh environments. But our adaptability may also be an evolutionary vestige: when our wandering ancestors crossed over mountain ranges, they had to adjust to the lower air pressure or die along the way. Can plants or other organisms that do not share our nomadic roots also adjust?

Amateurs can probe such mysteries of physiology thanks to Stephen P. Hansen, an innovative vacuum specialist in Amherst, New Hampshire. Many science enthusiasts already know Hansen from *the Bell Jar*, his quarterly journal devoted to amateur experimentation with vacuum techniques. Hansen developed a small and inexpensive chamber to conduct biological experiments at simulated high altitudes. Although technical complexities and ethical concerns would make it problematic to place one's pet hamster inside, this chamber is ideal for amateur investigations of less complicated organisms. Bacteria, insects, or small plants, for example, serve as ideal test subjects.

Hansen's device consists of a stainless-steel mixing bowl and a Pyrex bowl. The Pyrex bowl makes a perfect see-through top for the chamber, and the stainless-steel bottom allows easy installation of vacuum ports and electronic sensors.

You form the airtight seal by pressing the bowls lip to lip into a gasket cut from a thin sheet of rubber. Hansen secured his 12-inch (30.5-centimeter) metal bowl from United States Plastics Corporation (800-537-9724 or 419-228-2242; catalog number 84104). He bought a sheet of rubber from his local hardware store. You should be able to purchase a matching Pyrex bowl from just about any housewares merchant. Drill two holes in opposite sides of the steel bowl and epoxy ¼-inch (5-millimeter) brass hose barbs into each. You can obtain such fittings from a well-stocked hardware or plumbing supplier.

Although just about any mechanical vacuum pump will effectively draw down the internal pressure, the high price of most models will also draw down your budget. But if you restrict your research to terrestrial conditions—pressures no less than those at the top of Mount Everest—you can get by with an inexpensive type. Hansen used a surplus dry-vane vacuum pump, which he procured from C&H Sales in Pasadena, California (800-325-9465 or 626-796-2628; catalog number PC9703), for under $50. Regular vinyl tubing, affixed with steel hose clamps, makes fine vacuum line *[see Figure 4.1 on page 28]*.

For experiments restricted to simulated altitudes no greater than 4,600 meters (about 15,000 feet), a pocket altimeter is the best choice to monitor the pressure. Edmund Scientific in Tonawanda, New York (800-728-6999), sells one with a zero adjust for about $35 (catalog number 34-544). Just place the altimeter inside the chamber and read the equivalent elevation by looking at the scale through the Pyrex bowl. To mimic higher altitudes, you will need to hook up a vacuum pressure gauge *[see Figure 4.2 on page 29]*; a Bourdon gauge would do. These units, which you can buy at an auto supply store, read in either millimeters or inches of mercury below ambient pressure. Figure 4.2 shows the corresponding altitude.

Probably the most challenging problem you will encounter is regulating the pressure in your chamber. You could install an electronic sensor and control circuit that triggers the vacuum pump whenever the pressure

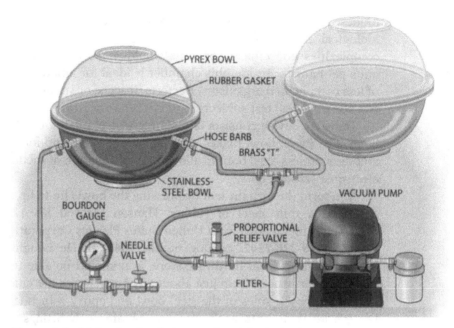

Figure 4.1 Mixing bowls of glass and metal attached to a vacuum pump make a spherical altitude chamber.

climbs above a chosen set point. But such equipment is difficult to build, and commercial systems cost hundreds of dollars.

Fortunately, cheaper solutions exist. Hansen found that an $8 proportional relief valve works quite well. His choice is model number VR25, manufactured by Control Devices in St. Louis, Missouri. (You can purchase one from W. W. Grainger: 773-586-0244; catalog number 5Z-763.) The vacuum pump pulls air from the chamber, and the valve allows a weak flow of air to pass in from the outside. This adjustable leak prevents the full force of the vacuum pump from acting on the chamber, allowing the internal pressure to stay higher than it would otherwise. The needle valve allows a weak but constant stream of air to flush out any gases produced by living things inside.

Although it is a bit tricky to adjust the proportional relief valve to a particular pressure, once properly set, the chamber will maintain that level reliably. Still, you will have to keep your vacuum pump running all the while. Note that if the outside air pressure rises or falls, so will the pressure inside your chamber.

Certain biological experiments might require you to vary the temperature and light levels. Try putting the chamber on a windowsill, resting the

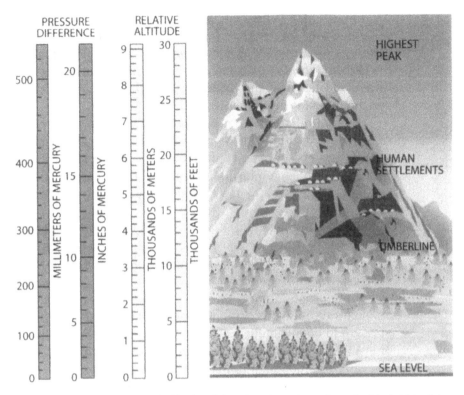

Figure 4.2 The simulated altitude of the chamber above its actual height is determined from the readings of a vacuum gauge.

stainless-steel bowl on a heating pad or submerging it in a bucket of ice water. For more precise adjustments, you may place the chamber under a full-spectrum light controlled by a timer and run a heat source from a thermostat. After struggling with homemade thermostats for years, I recently gave in and purchased a $165 unit from Omega Engineering (*www.omega. com*) in Stamford, Connecticut (800-826-6342 or 203-359-1660; catalog number CN8590).

If you go this route, you will also need to purchase 25 feet of type T thermocouple wire (Omega's catalog number is PR-T-24SLE-25). It consists of two wires made from copper and constantan laid side by side. Cut a length of this double wire so that it is long enough to reach from the controller to the inside of your chamber. From one end, strip two or three centimeters (about an inch) from the tips of both wires and twist the exposed leads together. Once intertwined, a tiny voltage develops that is related to the temperature of the coupled wires. Attach the other ends of the wires to the controller.

You will have to feed the thermocouple wires (and any other electrical cables required for your particular experiment) into your chamber without creating leaks. Another hose barb serves well here. Pass the wires through the barb and then secure them in place by filling the tube with epoxy. After the epoxy sets, install this assembly through the stainless-steel bowl just as you mounted the others earlier.

The controller I used can deliver a maximum of 5 amperes at 120 volts AC—that is, it is limited to a power output of 600 watts. Omega sells many suitable heaters, but you could probably coat the sides of the steel bowl with a sprayable foam insulation and rest the exposed metal bottom on a small hot plate, such as the kind designed to keep your morning coffee warm in the cup.

Finally, you may find it useful to build two identical chambers and tie them together with a T-shaped joint. This duplication will allow you to run two experiments simultaneously, using one chamber for tests and the other as the control. At times, you may want to keep the second chamber at full atmospheric pressure. But should you wish to vary light level or temperature only, this arrangement will let you maintain the same low pressure in both chambers. Try growing tropical plants at alpine altitudes or test whether a fly tires more quickly in thin air. With a little imagination, you can continue such rarefied pursuits indefinitely.

Amateurs interested in vacuum-related experiments can subscribe to *the Bell Jar.* Send a check or money order for $20 to *the Bell Jar,* 35 Windsor Drive, Amherst, NH 03031 (*http://www.tiac.net/users/shansen/belljar/*).

5 TIN-CAN KYMOGRAPH

by C. L. Stong, April 1960

O n one occasion or another most experimenters need a recording appa-ratus that can automatically plot the movement of a pen or a stylus against time. But commercial recorders are expensive. Norman D. Weis, an instructor at Casper College in Casper, Wyoming, was confronted by the problem while doing graduate work at the University of Colorado, and he decided to meet it head-on. He calls the result the tin-can kymograph. Basically Weis's instrument consists of a motor-driven drum mounted ver-tically on a thrust bearing and fitted with a sheet of smoked paper on which the graph is traced by a mechanically actuated stylus.

"The cost of this apparatus," writes Weis, "will vary from nothing to as much as $100, depending upon the builder's talents for adapting and scrounging. The secret of the low cost is found in the method of driving the drum that transports the record sheet. Instead of coupling a motor to the axle of the drum by means of gears, as is done in conventional designs, my drum is driven at its edge by frictional contact with an extension of the motor shaft. Moreover, when the apparatus is used to chart the respiration of a human subject, the stylus is actuated by a length of thread anchored by a block instead of the expensive chest-expansion tube and tambour-needle assembly familiar to students of biology *[see Figure 5.1 on page 32]*.

"The stylus assembly can be moved up and down the drum as desired, thus providing for several recordings on one sheet of paper. The number of recordings per sheet is limited only by the amplitude of the stylus excur-sions. Although normally actuated by a mechanical link in biological observations, the stylus may be coupled to any desired sensing device and driven pneumatically, electrically or otherwise.

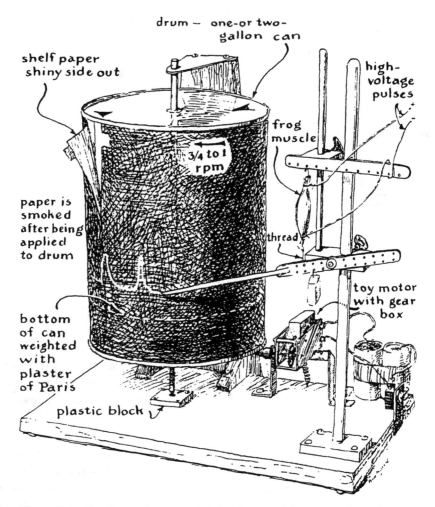

Figure 5.1   The tin-can kymograph set up to record the contractions of a frog muscle.

"A gallon can of the kind used for shipping fruit juice or syrup serves as the drum. The can is weighted with about an inch of plaster of Paris. An accurately centered hole ¼ inch in diameter is drilled through the top and bottom. One or more additional small holes in the top equalize the air pressure inside the can with that of the atmosphere. The drum turns on a shaft made of ¼-inch drill rod, or other straight material, inserted through the centered holes of the can and soldered in place. The ends of the shaft extend beyond the can 2 inches at the top and 1½ inches at the bottom. The bottom of the shaft should be cut square and smoothed with a fine stone. The drum assembly is supported by a thrust bearing: a piece of plastic ½ inch thick drilled with a centered hole ⅜ inch deep. A steel ball ¼ inch in diameter is placed in the bottom of the hole. The shaft turns on this ball. [Alternatively, you can purchase thrust bearings inexpensively from the McMaster Carr Company in Santa Fe Springs, CA. Call 562-692-5911 for their catalog or check them out online at *www.macmastercarr.com*. Ed.] The upper end of the shaft is supported laterally by a simple journal bearing: a hole through a piece of 16-gauge sheet metal screwed to a solidly braced column of plywood. The base to which the column, thrust bearing, and other components are screwed, is a piece of plywood ¾ inch thick, 12 inches wide and 16 inches long, finished with shellac. It rests on 4 rubber buttons, of the kind used on the bottom of chair legs. The drum assembly is removed from the instrument simply by lifting it from the thrust bearing, swinging the shaft to one side and sliding it out of the upper bearing.

"Any small motor will serve for the drive, but construction is simplified by using a motor with built-in reduction gears. Mine was taken from a mechanical toy that operated from a two-cell battery. Similar motors are sold by hobby shops and toy stores for use with model-construction kits. The drive shaft should turn at a rate of about 30 revolutions per minute. The shaft is fitted with a series of frictional rollers of increasing diameter. These are made of rubber tubing; a relatively long piece that makes a snug fit with the shaft is first pushed over the shaft. Progressively shorter lengths of increasing diameter are then telescoped over the first length. My rollers were made from three sizes of tubing: ⅛, ³⁄₁₆ and ¼ inch, as shown in the accompanying illustration *[see Figure 5.2]*.

"The active components of the stylus assembly include the stylus and its supporting lever arm together with a secondary fixture that is electrically insulated from the remainder of the apparatus. These parts are supported by a movable block that rides on and clamps to a vertical post attached to the base by a flange *[see Figure 5.3 on page 34]*. The post may be made of drill rod ⅜ inch in diameter. Any handy material, such as steel or plastic, may be used for the block.

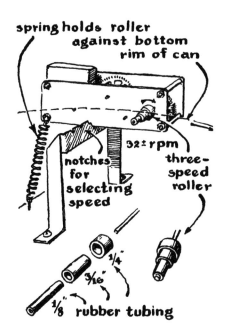

Figure 5.2 Detail of drive for kymograph.

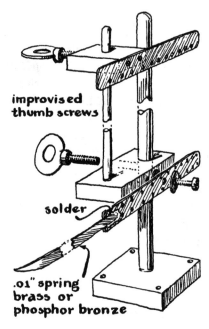

Figure 5.3 Detail of stylus for kymograph.

The lever arm of the stylus assembly is drilled in the middle and mounted on the block by a screw and washers. It must turn freely. The edges of the lever arm are drilled with holes ¾₂ inch in diameter spaced at ¼-inch intervals. These make it possible to balance the stylus by hooking a small weight to the bottom of the arm, and to attach specimens or apparatus to the top of the arm. A secondary post of ¼-inch stock extends from and above the movable block. This supports a second block made of insulating plastic, to which a metal strip is attached that will hold the upper end of a specimen. The lower edge of this strip is drilled with a set of holes to match those in the upper edge of the lever arm. The blocks are secured to their respective posts by thumbscrews. This arrangement permits the entire stylus assembly to be shifted vertically merely by loosening the thumbscrew that clamps the lower block.

"I find that recordings made on highly calendered (slick) shelf paper are quite sharp and easy to read. A single roll of paper lasts me for months. The paper is cut to match the depth of the cylinder and long enough to wrap around it with ¼-inch overlap. The direction of the wrap should be chosen so that the stylus slides off the top of the overlap—otherwise it may catch and tear the edge of the paper. The overlap is stuck together by bits of Scotch tape. The paper may be smoked over almost any flame deficient in oxygen: a Bunsen burner with the air supply closed, a candle or a kerosene lamp. [WARNING: There's a real fire danger here, so be careful. You must roll the paper on the can tightly before you expose it to the candle flame. Then the can will conduct heat away from the paper and prevent it from bursting into flames. Also, you must smoke the paper over a large sink. That way, should the paper catch on fire you can just let go of the can and quickly douse everything with water. Ed.]

"Graphs of respiration are made by stretching a thread over the chest of the subject, tying one end to the lever arm and anchoring the other under a book [see Figure 5.4]. (If the subject should sit up suddenly, the thread slides from beneath the book and spares the apparatus.) Variations

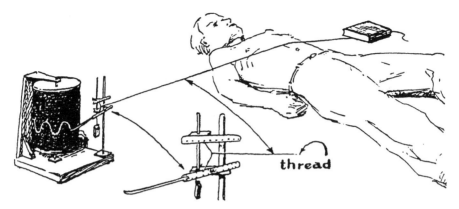

Figure 5.4   How the kymograph is used to chart respiration.

in respiration can be observed by having the subject rest for a few minutes prior to starting an experimental run or, conversely, by having him exercise vigorously immediately prior to the run. The same technique can be used to record any small physical movement." [Many experiments in biology will require you to record a slowly varying electrical signal. That's easy if you go to any store that sells surplus electronics and purchase an old galvanometer. These analog meters have a swinging arm that indicates the amount of current passing through the coil. To use this kymograph to record an electrical signal, just remove the outer casing of the meter and link the kymograph to the armature with a short length of thread. Ed.]

6 MICROSCOPY: SUBTLE SECRETS AND ADVANCED TECHNIQUES

by C. L. Stong, July 1955

The following primer focuses on advanced microscope techniques. You can use the information here to dramatically extend the capabilities of almost any modestly priced microscope. And since the microscope is such a fundamental tool to modern biology, the techniques described here can give the amateur biologist a huge boost. However, this article is not intended for the novice. And although I've tried to take the sting out of the jargon by adding an explanatory sidebar, even the advanced microscopist may want to read this chapter within arm's reach of a good text on optics. Also, the article assumes that you have a good medium grade microscope with a traveling stage, a color-corrected condenser, and a filter wheel beneath the stage to adjust the properties of the light that passed through your specimen. Ed.

The art of stage lighting has come a long way since the days of the showman P. T. Barnum and the microscopist Ernst Abbe. These pioneers knew that organisms, whether human or microscopic, rarely look their best in the head-on glare of white light. Both workers recognized the advantages of proper background illumination and learned to bring out modeling through the use of oblique light. The intervening decades have brought their techniques to a high state of development. Today's impresarios can multiply their effectiveness by mastering these techniques whether they preside over the stage of a theater or that of a microscope.

"When I started to use my first microscope," writes John De Haas, an amateur microscopist in New York City, "I thought the only way you could see significant detail in an animal of microscopic proportions was to kill and stain it. I was interested in studying protozoa, particularly the flagellates. But the application of make-up to my performers ended the show before the curtain went up. I wanted to study the internal structures of my animals as they swam, fed and reproduced. That meant that I had to work with light alone. As things developed, Rheinberg color illumination, dark-field lighting and even an inexpensive version of phase-contrast microscopy all proved within easy reach of my facilities. Astonishing increases in effective resolving power, I learned, are possible even with bright-field illumination. You simply block off 95 percent of the light from certain directions. Beginners often make the mistake of flooding their specimens with light and thereby washing out all detail.

"The effect is easy to demonstrate. Assume that you have a medium-priced microscope fitted with an eight-millimeter objective, an Abbe condenser and an eyepiece capable of giving the instrument a magnification of 150 diameters or so. [Note: A "condenser" is the apparatus that concentrates the light before it passes through the specimen and into the microscope. Typically, condensers are built directly into the microscope stage. Condensers with just a single lens perform poorly because they always refract shorter wavelengths at larger angles than the longer wavelengths. This results in a field of view that is redder in the center and bluer at the sides. By using more than one lens this so-called chromatic aberration can be at least partially corrected. An Abbe condenser is a simple and extremely popular variety that usually consists of two or three color-correcting lenses and an iris. Ed.] Put a slide of diatoms under the objective. Focus on one of them carefully and fiddle with the lamp, mirror and condenser adjustments until maximum detail appears. The chances are good that you will see the diatom in sharp outline against the bright field. The body will show little detail beyond a fuzzy pattern of striations. Substituting an eyepiece of higher power will not help much. The striations will appear bigger—but proportionately fuzzier. At this point many amateurs decide they need a better microscope. There is another way out, and it costs much less.

"Cut a disk of black cardboard to fit the filter carrier of the substage and make a quarter-inch hole in the disk midway between the center and the edge. Without disturbing the adjustment of the instrument, slip the disk into the substage. You will find a position where, despite the lower intensity, an astonishing amount of detail comes into view. The striations stand out sharp and clear and, depending upon the structure of the diatom, you will observe that the striations are rows of objects still more minute. By moving the disk slightly up and down or sideways, or rotating the stage, even smaller details can be resolved."

This technique, De Haas explained, is a form of oblique bright-field illumination. Nothing is coming though the eyepiece that was not there before. When the iris of the condenser is wide open, contrasting details of light and shadow in the image are submerged in glare because the deviated rays comprising the first- and lower-order spectra, that carry the fine details of the specimen, are faint when they are compared with the direct or zero-order rays.

It is interesting to examine and manipulate the spectra because this enables you to predict the resolution even before you look at a specimen. To view the spectra you observe the back lens by removing the eyepiece and substituting a pinhole. You will also need a series of opaque disks punched with a quarter-inch hole, the position of which progresses through the series from the center to the edge.

With the iris of the condenser wide open, focus the microscope on the diatom as described in the De Haas experiment and look through the pinhole at the back lens of the objective. The lens will be flooded with light to its edge. Now slowly close the iris. As the image of the iris in the back lens becomes smaller and smaller, the edges of two colored disks will appear at opposite sides of the field. The edges nearest the white disk will be an intense violet that shades into blue. These are images of first-order spectra. The fine structure of the diatom is acting as a diffraction grating, and hence some of the rays are deviated from the central beam. Now insert one of the opaque disks which has been perforated slightly off center into the filter holder and open the iris wide. The radius on which the perforation lies should be lined up with that of the spectra. The pinhole will now show that one of the colored images has in effect advanced into the field of view, while the white image and the other colored one have retreated an equal amount. Substitute in the filter holder the remaining opaque disks of the series one at a time for the initial disk. In effect this moves the perforation successively farther from the center. Observe that as the first-order spectral image advances more and more into the field, it shades through the colors of the rainbow and that, as it comes fully into view, a fainter companion appears at the edge. This companion is the second-order spectra. Depending upon the characteristics of your objective and the specimen, you may be able to entice as many as three orders into view before the white zero-order disappears. Moreover, by rotating the specimen or adjusting its lateral position slightly you may be able to bring other spectra into view at right angles to those already in the field. When you have accomplished this, substitute an eyepiece for the pinhole. The specimen will bristle with detail!

The explanation of why high resolution is associated with spectra is found in the wave nature of light. Except for shadows of gross objects cast by zero-order rays, the microscopic image is formed by interference at the focal plane of the eyepiece between rays that have been diffracted at the object.

The technique of increasing effective resolution by the use of an off-center diaphragm is a form of annular illumination, in which the specimen is lighted by a hollow cone of light. Most of the cone is missing in the above experiment, but it would be complete if a ring of perforations were extended all around the opaque disk. A luminous cone would then surround a solid cone of darkness. [Instead of perforations, use your computer and any graphic program to create a black disk with a thin clear circular band. Then use a laser printer to put this image onto a plastic transparency sheet. Place this in your filter wheel and you'll have a perfect cone of light illuminating your specimen. If you make several such filters, each with its circular band of a different radius, you'll be able to select the best angle for illumination for each specimen. Ed.] If the base of the dark cone were made wide enough, no direct light from the condenser would enter the objective; and in the absence of a specimen, the field would appear dark. A specimen would reflect some of the rays into the objective and thus cause it to stand out in bright contrast against the dark field. Such reflection is greatest at points where the refractive property of the specimen changes abruptly, as at its edges. Hence the usefulness of dark-field illumination is limited to specimens characterized by sharp contrasts in refractivity.

This is also true, for the same reason, of Rheinberg color illumination. Although Rheinberg illumination has limited value as a research tool, the fascinating results justify setting it up as an experiment. Essentially it is annular illumination in which an outer cone of colored light surrounds an inner cone of light in a contrasting color. As in the case of dark-field illumination, unless a special condenser is used, Rheinberg illumination works best with low-power objectives, preferably those of 16 millimeters or more used in combination with a 5-power eyepiece. To prepare a Rheinberg setup, cut a disk of colored gelatin or other plastic to fit the filter holder. Red is a good color for this disk. Perforate the center with a ½-inch hole. Next cut a ⅜-inch disk of contrasting color, say blue. Cement this disk over the perforation in the larger disk. The difference in size between the small disk and perforation allows for a ¹⁄₁₆-inch overlap. Insert the assembly in the filter holder and rack up the condenser until it is focused on the object plane. The field will now appear uniformly blue because the objective is immersed in the

A Note on Diffraction

This might help you understand what's going on. When light passes through something, like a specimen on a microscope stage it can both refract (bend in the medium) and diffract (scatter off a structure inside). Diffracted rays tend to fan out in circles and rays which are diffracted from different parts of the specimen interfere, that is, add together. Depending on how far apart these structures are (we're talking distances on order of a few wavelengths now) they can reinforce each other (which we call "constructive interference") or cancel each other out (which we call "destructive interference"). So a specimen viewed through just one color of light shows better contrast because these light and dark areas aren't washed out by light at other wavelengths. That's partially why good microscopes come with filters.

When a light ray diffracts, it spreads out in all directions into a spherical wave, which then interferes with other defracted waves. If the scattering points are very close together, these two wave trains are nearly identical and one has to look a distance of many wavelengths away from the scattering centers before one can see dark areas where the waves interfere destructively. The number of wavelengths away depends only on the distance between the scattering centers, but since each wavelength is a different physical size, this happens at different locations for different colors. As a result, when white light diffracts one sees a rainbow of colors with (the shorter wavelength) blue closer to the scattering centers than the (longer wavelength) red. These colors are called "spectra." Since the wave trains continue to spread out as the distance gets greater, the relative positions of the defracted waves continue to shift. The first dark band occurs when the waves are mismatched by one half of a wavelength. A second band occurs when they are mismatched by three halves a wavelength, a third dark band appears at five halves, and so on. The result is a repeating pattern of colors created by many regions of constructive and destructive interference.

About the jargon: Physicists use the word "phase" to describe the degree of mismatch between two interfering waves. Two waves that are shifted with respect to each other just enough so they interfere destructively (the peak of one wave intercepts the valley of another) are said to be 180 degrees out of phase. Note that shifting one wave by one complete wavelength corresponds to a phase shift of 360 degrees. So-called "zero-order" rays are those that pass straight through the specimen without defracting. The colors that are produced by waves that are mismatched by less then one half a wavelength (or alternatively, by two waves that are less than 180 degrees out of phase) are called "first order spectra." The second band of colors, those produced by waves that are mismatched by more than one half a wavelength (180 degrees out of phase) and by less than three halves (540 degrees out of phase), are called the "second order spectra." The third band of colors is the "third order spectra" and so on. Ed.]

central blue cone of light. Now place a small amount of water on a slide and drop in a grain of some effervescent substance such as Alka-Seltzer. The eyepiece will present a striking display as myriad bubbles bend red rays into the objective—like fireballs rising through a sea of blue.

Of greater interest to the advanced worker is the technique of phase-contrast microscopy, devised in 1934 by the Dutch physicist Frits Zernike. Phase contrast, like dark-field lighting, may be considered a form of annular illumination. Unlike other forms of annular illumination, however, Zernike's technique produces contrasts in the image by exploiting minute differences in phase between direct waves from the light source and those deviated by the specimen. The amplitude or brightness varies inversely with the square of the distance from the source, as suggested by the sine wave extending to the left from the lamp. The velocity of waves intercepted by the glass block is retarded by the refractive property of the glass and, in this case, the light emerges from the block a full wavelength behind that propagated through the surrounding air. Although both sets of waves are spherical, those that have traversed the glass block are now said to "lag" 360 degrees behind those of the source. The lens intercepts the spherical waves and, by virtue of its thickness increasing toward the center, retards the spherical wave fronts just enough so that they emerge as plane waves.

Many microorganisms are transparent and, like glass, can retard the velocity of light. Unfortunately the refractivity of many interesting ones nearly matches that of the medium in which they live. The eye is sensitive only to changes in amplitude or brightness. Hence microorganisms that cause only small differences in phase between light transmitted through them and through the surrounding material are invisible. Prior to the invention of phase microscopy they could be seen only after being stained or immersed in a fluid of substantially differing refractive index. These alternatives either killed them or seriously interfered with their natural processes.

The phase-contrast technique makes such objects visible by transforming small differences in phase into small differences in amplitude or brightness—to which the eye is sensitive. The trick is accomplished by retarding part of the light (passing it through a thin sheet of glass or similar material called a phase plate) so that all light arrives at the focal plane of the eyepiece in phase or 180 degrees out of phase. The crests and troughs of the waves are thus made to coincide or cancel. They "interfere," or combine their energies, and thus set up amplitude differences that constitute an image of the object.

The fine details of the image are carried by spectral orders of phase. These correspond to the spectral orders of amplitude that account for im-

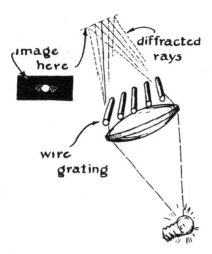

Figure 6.1 An amplitude diffraction grating.

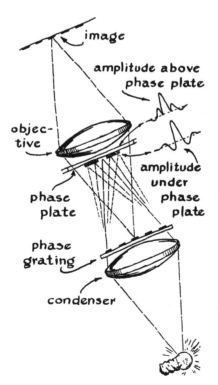

Figure 6.2 A phase diffraction grating.

age resolution in ordinary bright-field work, as demonstrated by the experiment in oblique lighting and illustrated in *Figure 6.1*. A phase grating, consisting of alternate strips of transparent material which differs slightly in refractivity from the intervening material, works much like the amplitude grating illustrated here. Transparent specimens can be considered phase gratings because the refractivity of their structure varies; rays transmitted through those portions of higher refractivity are deviated with respect to those transmitted by portions of lower refractivity. Thus two sets of waves enter the objective, distinguished only by their phase difference. It can be demonstrated mathematically that a third wave can be found which represents the phase difference between the direct and deviated ray. This difference wave is always just 90 degrees out of phase with the wave emerging from the specimen. [A 90 degree phase shift corresponds to a shift of one quarter of a wavelength. Ed.] A value of amplitude can be assigned to the difference wave so that when it is added to the wave deviated by the specimen, the sum equals the direct wave transmitted by the surrounding material. Zernike looked for the difference wave in nature—and found it! It is the phase spectra set up by the specimen and it carries the specimen's phase image.

As observed in the experiment with oblique lighting, the spectral orders spread across the aperture of the objective's back lens and converge at the plane of the eyepiece. The condenser is

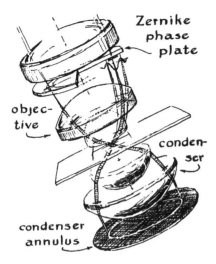

Zernike phase plate

objective

condenser

condenser annulus

Figure 6.3 The annular phase plate.

adjusted so that direct rays from the source meet at the back focal plane of the objective, as shown in *Figure 6.2*. Hence rays which pass through the center of the back focal plane and those transmitted through the complementary area (the spectral orders) are out of phase by 90 degrees. When this difference is adjusted so that the two arrive at the focal plane of the eyepiece precisely in or out of phase, interference takes place and an amplitude-contrast image results as in conventional bright-field illumination.

Zernike corrected the phase difference by inserting an annular phase plate in the path of the direct wave, as shown in *Figure 6.3*. In addition he coated the plate with a layer of light-absorbing material just dense enough to reduce the intensity of the direct wave to that of the deviated wave so that complete addition or cancellation would take place. This prevents the image from being masked by the excess light of the direct wave. As subsequently improved, plates are designed either to advance or in effect retard the direct wave and thus result in either constructive or destructive interference. Constructive interference causes a bright object to appear against a neutral background. Destructive interference reverses the effect. A set of phase plates, together with accessories including a condenser annulus and a telescope for aligning elements in the optical train, might sell for more than $1,000. De Haas devised a form of phase contrast for the amateur which approximates the results of the Zernike technique and costs about $100. It works well where phase difference amounts to a 20th of a wave or more.

De Haas's project was supported by a grant from the Pennsylvania Academy of Science and was reported in the Academy's journal. His setup requires a three-element Abbe condenser with the top lens removed, an iris or Davis diaphragm for use with a 4-millimeter objective of the dry type, a 16-millimeter objective, and individual stops (made with opaque lacquer on daylight filters of medium shade) for each objective *[see Figure 6.4 on page 44]*. These are inserted in the filter carrier beneath the substage condenser. The size of the stops must in general be determined experimentally. For a 16-millimeter objective of 0.25 numerical aperture the stop diameter

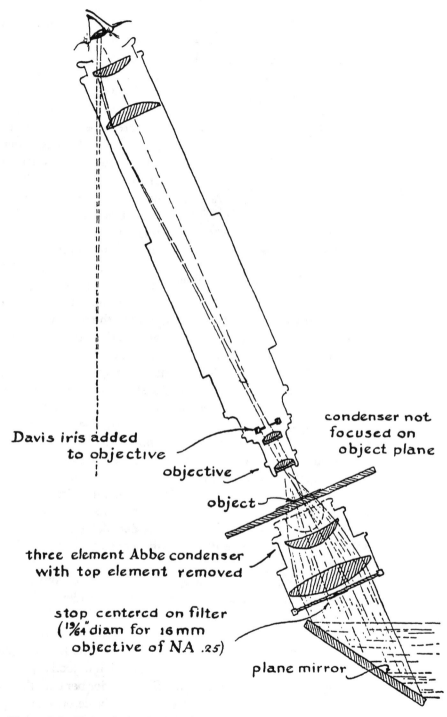

Davis iris added
to objective

objective

object

three element Abbe condenser
with top element removed

stop centered on filter
('¹⁹⁄₆₄"diam for 16mm
objective of NA .25)

condenser not
focused on
object plane

plane mirror

Figure 6.4 The optical train of the De Haas phase system.

should be $\frac{19}{64}$ inch. A 4-millimeter objective of numerical aperture 0.65 requires a stop of $\frac{21}{32}$ inch.

An appropriate stop is inserted in the filter holder, the eyepiece removed and the image of the stop carefully centered by eye in the back lens of the objective. The lamp must also be centered on the optical axis. A specimen is then placed on the stage and focused. A position of the condenser will then be found where details of the specimen stand out in sharp contrast—lighter or darker than the surrounding field depending upon the adjustment. When using the 4-millimeter objective, the position of the Davis iris influences the result and the best adjustment must be found by trial and error.

The technique contrasts with dark-field microscopy, where the condenser must lie in exact focus at the object plane and in which the object is always brighter than the field. The De Haas system appears to work on the principle of fringes diffracted by an edge—the stop providing the edge. The objective diaphragm probably acts on these fringes and those from the specimen so that they get added as in the Zernike system. De Haas has not attempted a theoretical analysis of the system.

PART 2

BOTANY

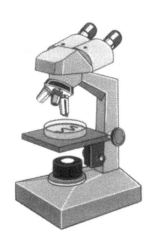

7 MUSEUM SECRETS FOR PRESERVING PLANTS

by Shawn Carlson, June 1999

The museum of natural history nearest you probably harbors an impressive collection of local plant life. Across the United States, these archives provide an excellent physical record that current and future biologists can use to track how native plants have fared in response to natural and human forces. Historically, amateurs have played a key role in shaping the botanical record, most notably since the Northwest expedition of Meriwether Lewis and William Clark, who preserved and returned scores of plants that were then unknown to science. Today each new summer brings an army of botanical enthusiasts scouring the countryside, searching for fascinating flora.

To aid budding botanists, I thought I'd share some museum tips for specimen preservation. You can use these techniques to help add to the official record or simply to engage your family in a rewarding outdoor adventure. Just don't run afoul of the law. Whether on private or public land, collect only if you have permission to do so from the authority responsible for the property.

Specimen preservation begins in the field. I suggest photographing each plant before cutting it to keep a record of it in its natural setting. Also, mark on photocopies of a topographical map the exact locations of your finds. Paste these sheets into your field notebook. If the plant is under 15 centimeters (6 inches) tall, collect the entire thing, roots and all. Otherwise, cut off a representative part, including flowers, fruits and any seed pods, which can often identify a plant better than its leaves. Tag each specimen with a small paper tab and record in your notebook the species' common name and scientific name if you know it, the date, and any details that a future botanist may need to know. Until you have finished your day's collecting, keep your cuttings hung upside down in the shade to minimize any crimping of their stalks as moisture begins to evaporate from their tissues.

Because cut plants deteriorate quickly, process them as soon as you get home. Begin by dipping each specimen in warm and slightly sudsy water, followed by gentle agitation in clean water to remove the soap. This process will kill bacteria and dislodge tiny crawlers. Thoroughly dry the foliage by blotting it with a paper towel.

Plants are best preserved by pressing and drying them. Begin by placing 3 layers of paper towels on top of a stiff board that measures about 30 by 45 centimeters (12 by 18 inches). Then gingerly lay out your cleaned plant, making sure to display different views (front and back) of its leaves. Large flowers should be split with a sharp knife and opened flat with their internal parts face up. Place 3 more layers of paper towels on top, followed by a sheet of corrugated cardboard and 3 additional layers of paper towels. Then lay out your next specimen. You can stack up to 10 cuttings this way [see Figure 7.1].

Place a second stiff board on top of the stack and apply steady, firm but gentle pressure to drive water out of the plant tissue and into the absorbent paper. Use a weight or four large C-clamps positioned near the corners. Or if you prefer, you can buy a professional press from a biological supply house. One of the largest is BioQuip (www.bioquip.com) in Gardena, California (310-324-0620; product no. 3115; $40). Or check out Fisher Science Education (*www.fisheredu.com*) in Burr Ridge, Illinois (800-955-1177; product no. CQS17670; $30).

Store your press on a warm, sunny windowsill. You'll need to refresh the paper every few days depending on how much water your specimens contain. Most cuttings do well with paper changes every two or three days, and they dry completely in about three weeks. But thick, fleshy leaves require daily replacements and can take four weeks to dry. Next, to kill any remaining tenacious pests, place the dried plants in a plastic bag and consign your collection to a freezer for at least three days.

Museum herbariums mount their specimens on cards measuring 29 by 42 centimeters (11½ by 16½ inches). BioQuip sells paper cards of that size for $1.70 per dozen (product no. 3135), and a buffered acid-free rag variety goes for $4.25 per dozen (product no. 3137). Fisher's price is $29 for 100 sheets (product no. CQS17676A). For those on a limited budget, ordinary card stock, though much smaller, works well and is available for under $10 in reams of 250 sheets from any office supply store. But if you

STIFF BOARD

PAPER TOWELS
AND CORRUGATED
CARDBOARD

Figure 7.1 The homemade press drives water out of plant tissue, preserving the specimens. The cuttings should be layered between paper towels and corrugated cardboard.

want your collection to be studied one day by botanists yet unborn, stay with acid-free paper.

Because dried plants are quite brittle, use extreme care when mounting them. Dilute some white glue by about one third with water and smear a thin layer onto a cookie sheet. Coat the backside of your specimen by gently settling it into the liquid. Delicately remove the plant, blot it on a sheet of newspaper and position it onto the mounting card. Dab all parts of your specimen with a paper towel to remove any excess glue. Place a sheet of wax paper and cardboard on top and use your plant press to secure the arrangement until the adhesive sets.

Although my early efforts using ordinary Elmer's white glue have held up nicely now for some 25 years, most professionals rely on a concoction they call "botany paste." BioQuip sells 2-ounce containers for a little over $3.

Transfer all the relevant information about each specimen from your field notebook to an acid-free paper label and glue it to the mounting sheet. Seeds and other loose parts can be stored by inserting them into thumb-size paper envelopes, known as fragment folders, which can then be glued or stapled to the sheet. You can easily make your own folders, or you can buy them precut from BioQuip in packages of 100 (product no. 3211BA; $15). And don't forget to include any photographs you took, which can be glued directly to the mounting cards. If the old adage is correct, each picture could save you a thousand words of exposition *[see Figure 7.2]*.

Last, you'll need to store your collection. My cuttings are organized inside loose-leaf picture albums that I keep inside two nested plastic trash bags. The specimens are contained within the innermost bag, which is

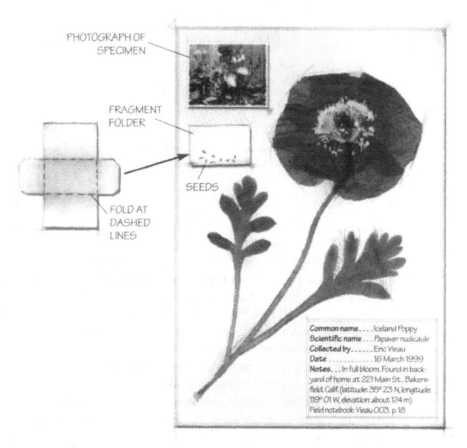

Figure 7.2 The mounting card may contain various materials and information, including photographs, seeds and field notes, in addition to the dried specimen itself.

tightly sealed. A fumigant bundle made of moth flakes wrapped in cheese-cloth sits inside the outer bag next to the opening of the inner bag. Changing the moth flakes every six months or so has kept away pests.

Living near an ocean allows me to collect sea plants. These organisms, however, present two special challenges. First, a plant that has washed up onto the beach is often long dead and is probably already home to thriving colonies of bacteria. But sea plants are quite tough and can tolerate rougher handling than their land-bound cousins. So, as soon as I get them home, I submerge them in hot and very soapy water for 10 minutes to suppress any bacteria.

The second problem is more subtle. Seaweed, if treated in the usual way, will rot. That's because the salt in its tissues absorbs moisture directly from the air. Thus, the plant remains perpetually wet. Fortunately, the salt can be leached easily away by a thorough soaking in distilled water. Pour into a basin at least 50 times more water by weight than the plant and let things sit for eight hours. Then do it all again. Adding a few drops of bleach each time will help keep new colonies of bacteria from taking hold while the salt diffuses out of the cells.

Once disinfected and thoroughly leached, seaweed can be pressed like any other plant. Rather than spreading out the foliage by hand, however, try arranging the plant while it is still floating in the basin. Gently scoop a sheet of card stock underneath the seaweed and carefully bring them both out of the water together. This technique captures the plant's natural motion, creating a more beautiful and realistic-looking specimen.

The author gratefully acknowledges informative conversations with Judy Gibson of the San Diego Natural History Museum.

8 GROWING PLANTS IN A CONTROLLED ENVIRONMENT

by C. L. Stong, March 1967

A miniature greenhouse with artificial sunshine and a controlled atmosphere can be built by the horticulturist for experiments with plants. The experimenter can regulate the color, intensity and duration of the light and the composition, temperature and humidity of the air. Hence anyone can rear plants under conditions that match those of almost any region of the earth in any season.

The miniature artificial greenhouse is simple and effective and so has found its way not only into the horticultural laboratory but also into the home of the city dweller who likes to raise unusual plants. One such person is Francis C. Hall, a lighting engineer who lives in Brooklyn, New York. Hall started after he had unexpected success with a single philodendron; now apartment gardening is an active avocation. He has built several miniature greenhouses, making as he did so some interesting innovations. He writes:

"After many attempts to grow flowering plants in our living room, I finally gave up. They all turned yellow, lost their leaves and died. So, like many of our neighbors, I settled for a wall pot of philodendron. This vine will live for a time almost anywhere, and it is easy to replace when it dies. Ours was replaced about every two months for a year or so until one of them unaccountably thrived, growing so vigorously that the tip of the vine reached the floor. I thought we had somehow acquired an exceptionally hardy specimen.

"At about that time we had a visit from a friend who teaches botany at a nearby high school. I called her attention to the thriving plant and asked for an explanation. After examining the pot and its surroundings she asked: 'How long has this table lamp been close to the vine?' I told her we had bought the lamp at about the time we planted the vigorous philodendron. The lamp is turned on every evening at dusk and stays on all night. 'There's your explanation,' she said. 'In effect you have converted this corner of your living room into a miniature greenhouse—not a very good one, but a greenhouse nonetheless.' She went on to explain the lighting requirements of plants and how I could improve my setup. I have been dabbling in amateur botany ever since.

"The key to a successful indoor greenhouse is the lighting. A combination of regular cool white fluorescent bulbs and incandescent lamps that are operated at about 95 percent of their rated voltage makes an effective combination. Neither kind of lamp will suffice alone because neither simultaneously radiates the copious amount of energy plants require in both the red and blue portions of the spectrum.

"Fluorescent lamps emit energy mainly in the violet, blue and green portions of the spectrum, and incandescent lamps emit mainly in the red and progressively less toward the violet. Experiments have shown that most plants require only the blue, red and far-red rays—the wavelengths extending from about 4,000 to 5,000 angstrom units in the blue and 6,300 to 7,500 angstroms in the red and far red. The mixture of greens, yellows and oranges that the eye perceives as a single shade of green are largely reflected by the plant pigments. Ultraviolet rays are damaging to plants, but most of the ultraviolet radiation in sunlight is absorbed by the atmosphere. The colors in sunlight that are essential to plant growth can therefore be provided by a properly balanced and filtered combination of fluorescent and incandescent lamps.

"In what proportions should the lamps be combined? When I built my first battery of lamps, the optimum proportion was thought to be 10 watts of fluorescent light to each watt of incandescent light. Since then the recommended ratio has changed. Some experiment stations have used ratios in which the wattages of fluorescent and incandescent light are equal. The results of my own experiments appear to favor the closer proportions.

"Power consumption is doubtless a poor index of spectral intensity because it neglects the performance characteristics of the lamps. A lamp of one type may radiate far more energy in certain portions of the spectrum than a lamp of another, depending on the design of the lamp and the voltage at which it operates. The variation occurs particularly in incandescent lamps. For fluorescent lamps I now use 30-watt cool white tubes of the rapid-start type equipped with ballasts, or transformers, designed to operate 40-watt tubes. Driving the lamps above their normal rating increases the emission of the desired blue light by 30 percent but does not appreciably shorten the life of the tubes. The trick will work only with 30-watt tubes.

"Ordinary incandescent lamps can be used for radiating the essential red light. When they are operated at their rated voltage, however, they do not emit strongly in the far red. For this reason I have switched to lamps that are rated at 130 volts. These lamps are used chiefly by industry in inaccessible places where lighting requirements are not severe and where the cost of replacing bulbs must be minimized. Although the lamps are rated at 130 volts, they are operated on 120 volts. Hence they last several times longer than conventional lamps. The yellowish-red emission is ideal for plants.

"These extended-service lamps are available from dealers in electrical supplies in the same wattages and at the same prices as standard lamps. They should not be confused with the long-life bulbs that are currently advertised. I now use six 30-watt fluorescent tubes in combination with six 15-watt incandescent lamps, a power ratio of 180 watts of fluorescent lighting to 90 watts of incandescent lighting.

[These days the home horticulturist can buy special fluorescent lamps called grow lights. These reproduce the sun's spectrum so well that plants have a hard time telling the difference. You can purchase them at any well-stocked plant nursery or at a shop that specializes in lightbulbs. Ed.]

"The structural details of mounting the lights and associated fixtures are determined largely by the application. For example, if the experimenter merely wishes to floodlight a shelf of plants, the hardware can be screwed to a simple wood frame suspended from the ceiling. In this case the fluorescent tubes can be mounted in standard twin-tube fixtures attached to the wood frame. The incandescent lamps can be spaced uniformly between the fluorescent fixtures. Normally the lamps will face downward. Heated air will rise from the lamps, so that porcelain sockets should be used. [Note, since grow lights produce very little heat, such extra precautions aren't necessary. Ed.]

"The optimum quantity of light to be used varies with the requirements of the plant. In nature plants grow under a wide range of light intensities, from as little as 10 footcandles to about 10,000. Philodendron will thrive at intensities as low as 50 to 100 footcandles; African violets do nicely at 600 footcandles, and orchids need 1,000. My experiments indicate that most popular varieties of flowering houseplants grow best at intensities ranging from 1,000 to 2,000 footcandles. Some growth chambers found in horticultural laboratories are equipped for lighting levels as high as 8,500 footcandles.

"Inexpensive light meters calibrated in footcandles are available for measuring the intensity. Alternatively, simple tables for converting the indication of photographic exposure meters to footcandles can be compiled easily. Intensity can also be estimated. A pair of cool white 30-watt rapid-start fluorescent tubes that operate with a 40-watt ballast in a twin fixture will deliver about 1,100 footcandles at a distance of six inches from the tubes, 650 footcandles at 12 inches and 500 footcandles at 18 inches. Within a 30-degree cone below the tubes the intensity falls off uniformly to about 80 percent at the edge.

"At present I have two small greenhouses. One, a lean-to, is installed outside a rear window of our ground-floor apartment. During part of the day the unit receives direct sunlight. The natural light is supplemented as desired by a battery of electric lights controlled automatically by a clock timer.

"The second chamber was constructed in the form of an ornamental cabinet, the basic details of which are shown in the accompanying illustration *[Figure 8.1]*. It consists of two compartments for growing stacked above a third compartment that houses an air conditioner, electrical controls, and miscellaneous supplies. The lamps are suspended from the inner surface of the top of the cabinet and the bottom of the top compartment. They are controlled by the preset clock timer.

"Lamps develop a substantial quantity of heat that must be removed to prevent the temperature of the chamber from rising above 75 degrees Fahrenheit, the maximum to which common houseplants should be exposed. Fluorescent tubes operate at relatively low temperature and present no problem. The associated ballasts radiate a fair amount of heat, however, and must be located outside the growth chamber. I installed mine on the top of the cabinet.

"Incandescent bulbs of the recommended industrial type are much cooler than conventional bulbs; they are warm rather than hot to the touch. Even so, the temperature inside a

ballasts

24"

air conditioner

top, bottom, shelves and back of 18-ga. metal

Figure 8.1 Francis C. Hall's indoor greenhouse.

glass and wood enclosure of 40 cubic feet equipped with 500 watts of mixed lamps will rise as much as 40 degrees F. above the temperature of the room. In order to remove this heat my cabinet was equipped with a small air conditioner of the type designed for window mounting (RCA Whirlpool, 4,700 British thermal units). *[See Figure 8.2.]*

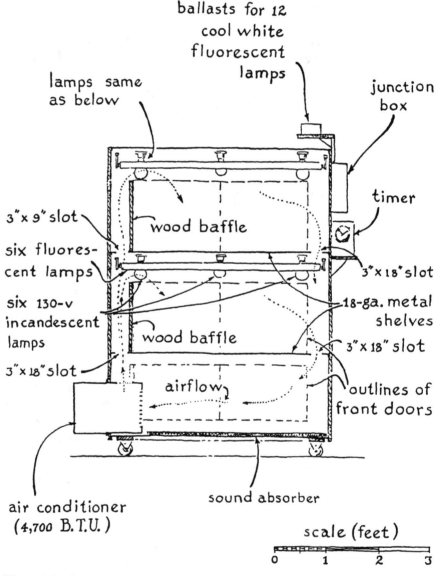

Figure 8.2   Details of the indoor greenhouse.

"The installation of automatic temperature controls adds considerably to the versatility of a growth chamber because plants require a daily rhythm of temperature change. This thermoperiod is analogous to the daily alternation of light and darkness. Experiments indicate that for most houseplants the night temperature should be allowed to fall about eight to 14 degrees F. below the daytime temperature. With geraniums I maintain a temperature of 67 degrees during the day and a temperature of 55 degrees at night.

"The control of relative humidity is difficult, and I have not yet succeeded in improvising an automatic system. My cabinets contain trays of moist gravel, and I water the potted soil periodically. Nonetheless, the relative humidity of the air tends to fall substantially on dry days. For this reason I measure the humidity with a psychrometer and spray the plants with water as necessary by means of a hand atomizer. On days when the relative humidity of the room air is high I depend on the air conditioner to remove the excess from the cabinet. Houseplants appear to do well at a relative humidity of between 50 and 80 percent.

"The length of the simulated day is the easiest variable to control. All it requires is setting a clock timer. The importance of the daily rhythm of light and darkness to the growth of plants was first reported in 1920 by Department of Agriculture botanists who were investigating the flowering of tobacco plants. Subsequently it has been learned that the photoperiod acts as a kind of trigger that determines when a plant will blossom, when seeds will germinate, when bulbs will form, and so on. The photoperiod also influences the color and size of leaves and the elongation and branching of stems. Commercial growers of flowers such as chrysanthemums routinely delay the flowering of plants grown in the field by switching on batteries of incandescent lamps for intervals as short as 10 minutes during the night, thus synchronizing the production of flowers with the demands of the market. I follow the same procedure when growing entries for our local flower show.

"What is the optimum photoperiod? The answer depends on the plant. In general plants can be grouped according to their preference for short, intermediate or long days. The perennial chrysanthemum and the poinsettia do best when the days are short—10 hours of light or less. Such a period is characteristic of plants that flower in the fall. Plants that do well at the opposite extreme—20 hours or more of daylight—include the China aster, the African violet, the tuberous begonia, and the philodendron. The third category, which includes the rose and the carnation, consists of plants showing little sensitivity to the photoperiod.

"Most of the popular houseplants with which I have experimented seem to do well on a daily exposure of 16 hours to light from which the ultraviolet rays have been filtered. This emission in fluorescent tubes occurs between 3,500 and 4,000 angstroms. It can be suppressed by inserting a sheet of Mylar W-2 plastic between the lamps and the plants or, if this arrangement is inconvenient, by wrapping each tube in a single sheet of the material.

"Experiments indicate that the triggering effects of the photoperiod are confined to the red rays in the vicinity of 6,600 angstroms. A few minutes of exposure to red light initiates biochemical reactions that continue for some time in the dark. Ordinary incandescent lamps emit enough red light to trigger the reaction in the case of houseplants. For example, houseplants that grow reasonably well on a window shelf will usually show dramatic improvement, particularly in winter, if they are grouped under a table lamp every day from dusk until bedtime.

"Like animals, plants must have food and water, each plant according to its needs. Conventional techniques of feeding and watering can be used for plants grown under electric lights. It is also possible to use the greenhouse for controlled experiments on nutrients. For instance, I once read that growth had been accelerated as much as threefold by fertilizing plants with carbon dioxide. The author went on to explain that, according to one hypothesis, the lush growth of plants during the Carboniferous period 300 million years ago resulted from the relatively large amount of carbon dioxide then present in the atmosphere and that the period ended when the carbon became locked up in deposits now represented by the fossil fuels. Why not flood my miniature greenhouse with carbon dioxide and grow giant plants?

"I quickly learned that the cost of the gas is too high for my pocketbook. It occurred to me, however, that something of the same effect might be observed if I sprayed the plants with carbonated water. I tried doing so but had poor results with bottled soda water. Apparently chemicals added to this water for retarding the escape of gas harm the plants.

"Carbonated water can be made at home, however, by means of a special bottle that accepts gas from small metal tubes. The homemade product worked. After trying various methods of applying the water in varying amounts I learned that a light mist applied to the leaves every other day doubles the growth rate and reduces the time required for the plant to reach maturity. [I must add that you can very inexpensively manufacture all the carbon dioxide you'll need by mixing baking soda with vinegar. Ed.]

"The relatively small amount of gas liberated from the water does not substantially alter the ratio of carbon dioxide present in the air of the chamber. Why, then, do the plants respond so dramatically? I can only guess that the leaves absorb the gas through their stomata, or pores. The application must be made with an atomizer that develops a fine mist, and the leaves should be moistened only lightly. Moreover, the relative humidity of the chamber should be measured after the treatment and lowered if it rises excessively. Some experimenters who tried the procedure without initial success made the mistake of ignoring the relative humidity.

"Otherwise my plants receive conventional fertilizers applied according to the established requirements of each species. Much annoyance and expense can be avoided by using hygienic methods. Pans of water kept on the shelves for maintaining humidity encourage the growth of algae and fungi. The pans should be removed and cleaned periodically when the plants are washed. Much unwanted growth can be discouraged by adding an algaecide to the water in the humidifying pans. I use a solution that consists of one ounce of cupric sulfate in one quart of tap water. One fluid ounce of this stock solution is added to each gallon of water used in the pans. Water so treated must not be allowed to come in direct contact with the potted soil. To prevent such accidental contamination surround the pots with plastic liners.

"Seedlings can be developed easily in the greenhouse. They do best in blue and red light; far-red light retards them. To make seedlings sprout I remove all industrial incandescent lamps and substitute a single standard 15-watt lamp. The result is a 12:1 ratio of fluorescent to incandescent light; this minimizes emission of the far red.

"Seedlings require a somewhat higher temperature for maximum rate of growth than mature plants do. The temperature should be between 75 and 80 degrees F. The soil can often be warmed to this temperature if the plants are raised to within four or five inches of the lamps. Alternatively, the added heat can be developed electrically by installing heating cables in or under the trays. Cables specially designed for this purpose can be bought from dealers in gardening supplies. It is good practice to plant seeds in sterilized soil in order to discourage the growth of fungi and molds.

"The cost of my greenhouse, including the air conditioner, was about $300. I can easily imagine a more elaborate and more costly installation. On the other hand, it is possible to conduct many fascinating experiments with little more than a potted plant and a single incandescent lamp. In my opinion few hobbies return more in terms of satisfaction per dollar and none appears to be attracting enthusiasts more rapidly."

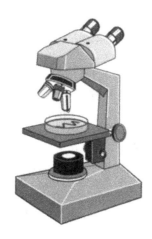

9 THE ESSENCE OF HYDROPONICS

by Albert G. Ingalls, October 1952

Gardening probably claims the interest of more enthusiasts than all other avocations combined. Whether the plot comprises acres of winding paths and formal beds or only a bit of potted soil on the ledge of an apartment window, those who do the tilling share a common love of things that grow. Whether they know it or not, gardeners are confirmed experimentalists, everlastingly testing nutrient, location, lighting, and variety. To the extent that these efforts are guided by observation, analysis, and test, the gardener is also an amateur scientist.

Nat E. Mankin of Chicago, a railroad freight-expediter for a nationwide transportation concern, is an extreme example of the casual gardener turned amateur botanist. His work in the field of soilless gardening, or hydroponics, has attracted wide attention in both professional and lay circles. His experimental techniques rank with those of the best professionals. Despite the impressiveness of his accomplishments Mankin's two decades of off-hour fun have cost him little in terms of time and money—no more than other city dwellers spend maintaining a few potted plants. "A few seeds and $10 worth of chemicals," he says, "will keep you going a lifetime."

Experiments with growing plants in liquid solutions, according to Mankin, date back to the days of the Roman Empire, when plants were grown in jars and vessels to which fertilizer was added from time to time. History does not indicate what objective prompted the experiments of the Romans, nor how their scientific studies, if any, came out. But in view of the complex chemistry of organic fertilizers and the wide variation of their quality, it is safe to assume this early work lacked any form of experimental control. The first recorded attempt at a controlled study appears to have been made in 1699 by an English botanist named Woodward. He grew spearmint in water containing an extract of soil. Although he observed that without aeration plants continuously immersed in a solution soon turn yellow and die, Woodward did not achieve true hydroponic culture because he, and everyone else, had insufficient knowledge of chemistry.

The credit for pioneering hydroponics, in the modern meaning of the term, goes to the French chemist Jean Boussingault, whose work in South America during the early decades of the nineteenth century proved that plants could be grown in sand, charcoal, quartz and other inert materials to which inorganic solutions were added. His studies proved that plants cannot assimilate free nitrogen from the atmosphere, and he first evaluated the role of manures. In recognition of these and related contributions, he was invited to occupy the chair of chemistry at Lyons, moving later to that of agricultural chemistry in Paris.

Boussingault's methods were quickly taken up independently by two German workers, A. Knop and Julius von Sachs, the latter a botanist at the University of Würzburg. They first focused attention on the fact that growing plants operate like the world's largest chemical industry, and that with the proper control plants are powerful tools for determining how nature converts the simple chemicals of the atmosphere into complex food substances. The controlled techniques of Knop and Sachs for studying the irregularities affecting soil-grown crops have survived nearly a century of use; their basic formulas for nutrient solutions still serve as the starting point for most hydroponic research.

Despite its long history hydroponics did not become a popular branch of amateur science until 1929. In that year William F. Gericke, a plant physiologist at the University of California, developed a special technique for applying hydroponics to commercial crop production. The transition from laboratory tool to business enterprise was attended by widespread publicity. Soon seed stores from coast to coast were sending customers from their counters laden with packages of mineral salts.

"The successful growth of a robust crop," says Mankin, "is based on the balanced supply of two classes of nutrients: the major fertilizing element which supply most of the plant's food requirements, nitrogen, potassium, calcium, phosphorus and sulfur, and the minor trace elements such as iron, manganese, boron, zinc and copper. Minute amounts of these trace elements, along with vitamin B_1, play a decisive role in maintaining the health of plants.

"The precise function of all the trace elements is not known—one of the things that attracts both professional and amateur botanists to the hydroponic technique. Without iron no green coloring forms in the plant. In the absence of boron no seeds develop. Deficiencies in other trace elements cause plants to be stunted, lacking in color, or malformed. In contrast, a well-balanced diet of major and minor elements makes it possible to grow plants more luxuriant than almost any that can be grown in soil. By soilless culture a single tomato plant of the Marglobe variety, for example, can be grown to a height of 25 feet and made to bear 20 pounds of perfect tomatoes. So effective is the technique that the Army Air Force maintained soilless gardens on Ascension Island in British Guiana and on Iwo Jima during World War II for the large-scale production of food crops. A number of commercial enterprises with personnel quartered in the Tropics and arid parts of the world operate similar installations."

Soilless culture takes one of three forms, according to the means of mechanically supporting the plants: sand, gravel or water. Each of the three techniques has certain advantages and disadvantages, but all share the common distinction of giving the experimenter more control of the plant's nutrition than is possible with soil.

Sand culture, developed for commercial purposes by New Jersey and Rhode Island agricultural experiment stations during the 1920s, is perhaps the simplest of the three methods. Mankin recommends it highly for the beginner. "Seeds planted in a sand-filled flower pot," he explains, "are kept moistened with nutrient solution until they mature as full-grown plants. Like the soil of the conventional garden, the sand must have drainage, and it should be flushed with pure water about once a week to remove excess mineral salts. This technique is also known as 'slop' culture. Its advantage lies in the fact that fresh air is brought into contact with the root system daily when the application of the nutrient solution drives stale air from the sand. The beginner may expect gratifying results from sand culture, at least equal to those he can achieve with a good quality of topsoil. The principal disadvantage of sand culture lies in the fact that the nutrients in a solution tend to crystallize on the grains of sand. This obviously alters the concentration of the nutrients, and deprives the experimenter of precise chemical control over his culture. The variation in nutrients may seem slight when considered in absolute terms of grams or ounces, yet it is sufficient to induce astonishing changes in a plant's metabolism. The presence or absence of one part of vitamin B_1 in a billion can make the difference between a healthy plant and one that is seriously ill. Another disadvantage of the method is that many common sands contain soluble minerals that contaminate the nutrient solution or radically alter the concentration of its trace elements. Sand need not contain much ferrous sulfate to alter the concentration of a nutrient solution calling for one part of iron per million."

For commercial soilless culture, gravel has several practical advantages over sand or water. Chief among these is the relative ease and speed with which solutions may be pumped into and out of gravel-filled growing-tanks. The medium may consist of any coarse-grained, chemically inert solid, ranging from stream gravel to crushed granite and coal cinders. As with sand culture the medium tends to introduce variations in the nutrient solution. Gravel culture is less exacting than water culture, but if the experimenter fully exploits its conveniences it requires almost as much equipment. Mankin urges the beginner to master the sand technique, and then to shift directly to water.

"Amateurs who are reasonably handy with tools," writes Mankin, "can build their own growing-tank for experimenting with water culture or 'pure' hydroponics. A common 5-gallon plastic bucket can with little labor be converted into a serviceable tank. The tank is first fitted with a growing-tray. This is made of conventional 1- by 4-inch stock of pine, fir, spruce, or white cedar. Redwood is toxic to most plants. The tray should be fitted to drop easily into the tank. The corners may be joined with corrugated fasteners, but for necessary rigidity and strength they should be reinforced with wood screws.

"The bottom of the tray is covered with either a fine-mesh chicken wire or a hardware cloth of heavy gauge: ½- to ¾-inch mesh. The mesh is attached to the tray by staples, and is supported by two or three narrow widths of metal strip spaced evenly across the bottom. The netting must receive a protective coat of enamel-based paint. The tray is suspended in the tank by means of three metal angle-braces extending from its upper edge. Finally the netting is covered with an inch or so of shredded excelsior and, over this, enough sphagnum moss, glass wool, dried hay or even coarse sawdust to fill the tray to its top. Make sure that no shreds of this litter extend through the mesh or over the sides. Toxic materials such as redwood sawdust should be avoided *[see Figure 9.1 on page 66]*.

"For introductory work it is advisable that the amateur use young transplants whose roots have been gently washed free of soil. Tomatoes take readily to pure hydroponic culture, and hence they are good plants with which to gain experience. A hole is made through the litter to receive the plant; a sharpened stick will serve as a suitable tool. The plant is inserted through the litter so that about an inch of stalk separates the root system from the mesh. The plant is then secured in position by packing the hole with moss or other litter. Be careful not to injure the plant by applying heavy pressure, but close the hole completely. The solution must be kept in the dark to avoid the growth of algae.

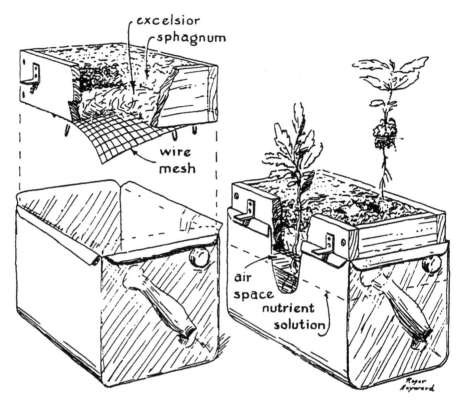

Figure 9.1 A homemade tank for the culture of plants in water.

"The root system must be supplied with oxygen; thus some means of aerating the solution must be provided. A piece of glass tubing closed at one end with a porous ceramic plug may be inserted through the litter so the plug rests against the bottom of the tank. This is coupled to an air pump through a length of rubber hose. Complete aeration rigs, including an electrically driven air pump, can be purchased for less than $50 in stores specializing in the needs of tropical fish, or the experimenter can do the job with a bicycle tire-pump. The aerator must run for at least 15 minutes twice a day. Another scheme that works well uses the so-called 'continuous flow' method of aeration. Solution drips continually from an elevated container into the center of a narrow-necked funnel. A length of tubing leads from the spout of the funnel to the bottom of the tank. Bubbles trapped between successive drops are carried into the tank, which must also be equipped with an overflow siphon. [See Figure 9.2.]

"Two alternatives are available with respect to preparing the nutrient solutions. Seed stores market a number of packaged plant foods. Simple

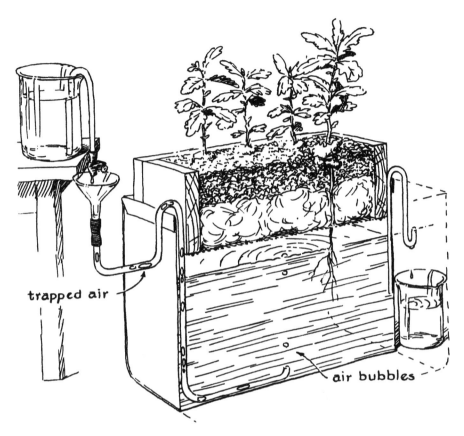

trapped air

air bubbles

Figure 9.2 The continuous-flow method of aerating a water-culture tank.

directions on the package explain how to make up the solution. The use of these preparations obviously limits the extent to which the amateur can manipulate the growth process. In contrast, solutions prepared from basic formulas encourage experiment and enable the advanced worker to vary the plant's nutritional intake at will. Unlike conventional gardening, water culture enables the experimenter to manufacture any variety of 'soil' desired.

"Enough solution should be prepared to fill the tank to within an inch and a half of the bottom of the tray. This air space between the top of the solution and the tray is important. From it the hairs near the top of the root system take up oxygen.

"All ingredients for a basic plant food are available through most drugstores. Just as no universal diet exists for all animals, no one formula meets the nutritional requirements of all plants. But plant physiologists in a number of universities and agricultural experiment stations have developed suitable nutrients for groups of common plants.

"The following formula will give good results with most common garden vegetables and household flowers:

MAJOR ELEMENTS

Compound	Grams per gallon
Magnesium sulfate	1.04
Monocalcium phosphate	0.54
Potassium nitrate	2.20
Calcium sulfate	3.04
Potassium chloride	1.60

MINOR ELEMENTS

Compound	Grams per gallon
Ferrous sulfate	.01
Manganese sulfate	.004
Boric acid	.0056
Copper sulfate	.0004
Zinc sulfate	.0004
Thiamine chloride	trace

[Other simple formulas that can be compounded by the amateur and modified to compensate for variations in the worker's local climate or to meet special research objectives may be found online. Ed.]

"All solutions must be adjusted for pH, or acidity. Most common vegetables and flowers prefer a slightly acid solution, their tolerance extending from pH 5 to 6 on a scale calibrated from 0 (extremely acid) to 14 (extremely alkaline). On this scale 7 represents a neutral solution. The experimenter must procure a kit for making pH tests. These are available through most seed stores. Although these kits take a variety of forms, most employ chemically treated paper, the color of which changes when it is dipped into a nutrient solution. The test paper is then matched against a varicolored chart calibrated in pH. The pH of the nutrient solution may then be adjusted to the correct range by adding minute amounts of sulfuric acid or potassium hydroxide. If the solution is too acid, potassium hydroxide is added; if it is too alkaline, acid is used."

Similar tests, described in handbooks on chemistry, enable the advanced amateur to maintain a close check on the concentration of all the minerals in the nutrient solution. Although these tests are useful, the most powerful indicator is the plant itself. By continuously observing and recording changes in the size, color, rate of growth and general health of the plant, its body, root system and leaves, the amateur learns to detect hunger signs that lead to refined growing techniques. By varying the proportion and amounts of the minerals, ideal foods can be developed for each species of plant in relation to its local ecology. The role of each element can be observed, and thus the art as well as the science of hydroponics can be mastered.

10 GEOTROPISM: THE EFFECTS OF GRAVITY ON PLANT GROWTH

by C. L. Stong, June 1970 and Shawn Carlson, February 1996

This article has special meaning for me. It describes one of my Grandpa Don's many amateur experiments and was originally published when I was just ten years old. My grandfather was so proud of this article, and he challenged me to see if I could improve upon his techniques. I did. And meeting his challenge (see the second article below) placed my feet firmly on the path of experimental science. I've never looked back.

The two experiments that follow have been the source of a great many amateur research projects, including several student projects that won high honors in national science fair competitions. I hope you enjoy them.

Ed.

Soon after a potted plant has been laid on its side the stem turns up and the roots turn down. Experiments indicate that such changes in the direction of growth are induced by gravity: plants tend to align themselves in the direction of the earth's gravitational field. Botanists refer to this tendency as geotropism and have discovered by experiment that it arises from the influence of gravity on certain substances in plants, namely the organic compounds known as auxins, which stimulate the growth of the upper parts of plants but appear under some conditions to suppress the growth of roots.

When a potted plant is inclined from the vertical, auxin concentrates in the lower sides of the stem and roots. The concentration causes the lower side of the stem to grow faster than the upper side. The stem bends upward. Conversely, auxin in the lower side of the roots retards growth, but normal growth continues in the upper side. The root turns downward.

These effects can be observed by making a simple experiment. With India ink draw a set of evenly spaced marks along the lower side of both the root and the stem of a seedling. Place the seedling horizontally in a moist container for 24 hours and then examine the marks. The spacing between the marks will have increased on the lower side of the stem where it bent upward but will not have increased on the lower side of the roots.

Of course this experiment does not prove that gravity is solely responsible for reshaping the plant. The tops of plants grow toward sources of light, and the leaves of many plants follow the sun. The unequal distribution of auxin is also responsible for this effect, which is known as phototropism, but botanists have not yet learned how light influences the substance. Three mechanisms have been suggested. Light may inhibit the production of auxin on the exposed side. Alternatively, it may denature a portion of the normal production. On the other hand, as in the case of gravity, light may cause auxin to move to the shady side of the stem. Whatever the mechanism, you can identify the side of the stem that grows faster by drawing evenly spaced rings of India ink around the stem of a potted plant, placing the pot upright near a window and measuring the spacing of the rings as the stem bends toward the light. This experiment casts doubt on the assumption that the stem of an inclined plant turns upward in response to gravity. Perhaps the stem is merely seeking light, which usually comes from above.

All doubt concerning the role of geotropism in plant growth can be resolved by another experiment that was first performed about 150 years ago. In this experiment upright pots that contain seeds or seedlings are uniformly illuminated on all sides, but the gravitational field is tilted from the vertical by mounting the pots upright on the rim of a wheel that turns in the horizontal plane. (Each pot is enclosed in a transparent container for protection against currents of air.) When the wheel turns, the pots are acted on by two components of inertial force: a horizontal component arising from the circular motion of the wheel and a vertical component resulting from the acceleration of gravity. The resultant force acts at an intermediate angle that is determined by the speed of the wheel.

The roots of plants that are grown on the continuously rotating wheel extend outward and downward at precisely the angle of the resultant force.

The stems grow inward and upward in exact alignment with the roots. The lines of resultant force along which the plants grow trace a cone in space as the wheel rotates. The altitude of the cone varies inversely with the speed of the wheel, an experimental result that can be explained only on the assumption that inertial force strongly influences the direction in which plants grow. Having established this fact, experimenters appear to have closed the books on geotropism and shifted their attention to other matters.

Interest in geotropism has been dramatically rekindled, at least for one amateur, by the development of space vehicles. About a year ago Don Graham, who is a commercial artist in Petrolia, California, began to wonder how a plant might react if it were grown in a weightless state. Graham decided to undertake the experiment but could think of no way of eliminating gravity without putting plants in a space vehicle. Instead he devised an apparatus that interferes with the natural response of auxin to the gravitational field. Plants that are grown in the apparatus apparently lose their sense of direction. Graham built a pot that rotates slowly but continuously in all coordinates of three-dimensional space and undertook to grow corn in it. He describes the experiment as follows:

"My apparatus consists essentially of a cylindrical pot that rotates simultaneously on its axis and in the horizontal plane *[see Figure 10.1]*. The hollow cylinder, made of the wire mesh known as hardware cloth, is a foot long and about four inches in diameter. The ends of the cylinder are closed by two wood disks. The cylinder is supported by a shaft that passes through snugly fitting holes in the center of the disks and is rotated on its axis by a pulley on one end. The shaft is supported at its ends by a pair of vertical brackets that are fastened to a horizontal wood base.

"The base is rotated in the horizontal plane by a vertical shaft that is coupled to a slow-speed motor by a belt. The motor turns at eight revolutions per minute. A 1:8 pulley ratio reduces the speed of the vertical shaft to one revolution per minute. The pulley that drives the cylinder is coupled by a belt to a fixed pulley attached to the frame on which the motor is mounted. None of the dimensions are critical, but the diameter of the fixed pulley should not be a multiple of the diameter of the driven pulley, because this ratio would generate a cyclical pattern of cylinder positions. A 7:11 ratio works well.

"The cylinder is filled with a mixture of four parts of sphagnum moss to six parts of rich loam. I moistened the soil and packed the cylinder as though it were an ordinary pot. Sweet corn was selected for the experiment because the seedlings of corn develop in the form of a series of concentric whorls that appear to be stronger and sturdier than most plants are during the first few days of germination.

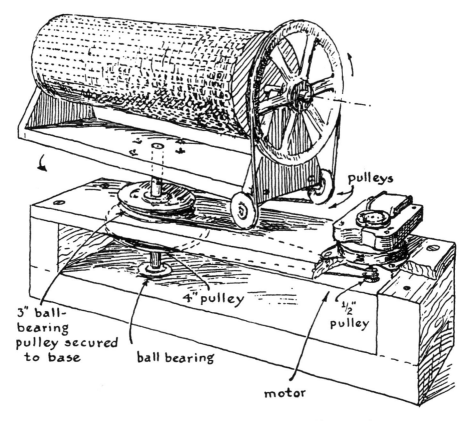

Figure 10.1 Don Graham's apparatus for experiments with geotropism.

"Seven uniformly spaced openings, each ½ inch square, were cut in the a wire mesh to form a helical path of one full turn that extends to within an inch of the ends of the cylinder. With tweezers I pushed a seed through each opening and into the soil to a depth of 2 inches, which is to say to the middle of the cylinder. The cylinder was wrapped with a single sheet of clear polyethylene to conserve moisture.

"The apparatus was placed on the ground in the backyard, where it would receive full sunlight. The motor was turned on and operated continuously for 14 days, except during brief intervals when it was stopped for a check on the temperature and moisture of the potted soil. During this entire period seven additional seeds of the same stock were growing in an adjacent garden area that contained identical soil. These plants served as controls.

"On the fourteenth day all seedlings (both the experimental ones and the controls) were removed from the soil, washed gently, measured, and replanted in the garden. The seven control seedlings had grown to an average height of 2½ inches and appeared to be normal in every respect. The most vigorous measured seven inches from the tip of the root to the tip of the longest leaf.

"The experimental seedlings had grown as vigorously as the controls. The largest measured nine inches from root tip to leaf tip. There the similarity ended. Whereas the controls grew straight up and down, most of the seedlings were sadly misshapen. Only one plant had found daylight; it grew about 2½ inches beyond the wire. The root, which was about 3½ inches long, bent randomly through the soil. One seedling grew in reverse: the root penetrated the wire and the stem remained in the soil. The root and stem of another seedling grew parallel in the same direction! One seed failed to germinate. Another produced a short root and an even shorter parallel stem. No experimental seedling had grown in the normal up-down direction. As the plants were removed I made a record of the direction in which each had grown with respect to its position in the cylinder. The record indicated that the direction of growth had been random *[see Figure 10.2]*.

"All seedlings matured in the garden, where they were cultivated and weeded regularly. The controls grew to heights ranging from 5 to 9 feet and yielded an average of 4 ears of corn per stalk. In contrast, the confused seedlings matured at a height of less than 4 feet. Only one experimental plant produced an ear, and it was a distorted, underdeveloped runt *[see Figure 10.3 on page 76]*.

"Other experiments involving geotropism come to mind. For example, how long can a germinating plant survive without damage in the absence of a normal gravitational field? My plants were rotated for 14 days. How much damage might have been evident if I had transplanted the seedlings after the fourth or the eighth day? How would a plant react to an increase or a decrease in the intensity of the gravitational field?

"I can think of no practical apparatus for lowering the strength of gravity on the earth to, say, that of the planet Mars. On the other hand, it is easy to investigate the influence on germinating seeds of an inertial force greater than the earth's gravity by growing plants on the rim of a wheel that is spinning. It might be interesting to find out how sweet corn would grow on Jupiter, where gravity at the surface is 2.6 times stronger than it is on the earth.

"One should not place too much confidence in the outcome of a single experiment. Nonetheless, having observed the reaction of my confused corn, I suspect that no plant in an advanced stage of evolution can grow normally in a weightless environment. Nor can such a plant reproduce

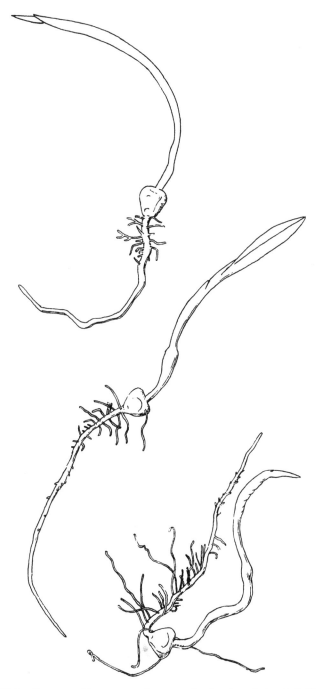

Figure 10.2 Corn seedlings grown in a constantly changing gravitational field.

Figure 10.3   Seed ear produced by an experimental plant.

itself for more than a few generations, notwithstanding the fact that one of mine developed seeds. Perhaps lower marine organisms such as algae, corals, or fungi could multiply in the absence of a gravitational field. So far as higher plants are concerned, however, gravity appears to be as essential to growth as sunlight. In my opinion, an orbiting spacecraft would make a poor garden."

My grandfather's article appeared when I was just ten years old. His comment near the end, that he could think of no practical way of lowering the strength of gravity to, say, that of the planet Mars, got me thinking. Two years later I built the bicycle wheel apparatus shown here that will simulate Martian gravity, or any gravitational field between zero and one g. Unfortunately, I never tried to publish it. After all, I thought, I was only an amateur botanist, and just 12 years old at that. My apparatus was so straightforward and the physics behind it so obvious that the method just had to be well known to professionals. So my invention gathered dust until 26 years later when I became a columnist for *Scientific American* and I needed a good amateur project to write about. When a description of this device finally appeared in print in February 1996 (and is reproduced below) I got calls from a number of excited NASA scientists who specialize in geotropism and who insisted that my technique was new to science. One even said that the article inspired him to start experimenting before the amateurs made all the discoveries!

There's an important moral here for all amateur scientists. Never assume away the value of your own creative work. You just might be the first person to put a good idea into practice. So get your inventions and discoveries out there! You can't make contributions to science until you do.

Ed.

Something remarkable happens when you tip a plant. Special hormones, called auxins, begin to collect in the underside of its roots and stem. Auxins stimulate stem cells to grow and divide. The bottom of the stem then outgrows the top, causing the stem to bend skyward. Auxins in the root cells act differently: they retard growth. The auxin-poor cells near the top of the root then outgrow the auxin-rich cells near the bottom, and the root bends downward. In this way, a tipped plant makes internal adjustments to realign itself with the pull of gravity.

Botanists call a plant's response to gravity geotropism. In the early 1800s experimenters explored geotropism by growing plants on a rotating wheel, thereby exposing them to both the earth's gravity and centrifugal forces. The plants grew against the direction of the combined forces.

But scientists soon learned that plants respond to gravity only sluggishly. Most plants must be tipped for at least a minute before the auxins start to redistribute. By the turn of the century, scientists had invented a device, called a clinostat, that tricks plants into thinking they are growing in near-zero-gravity environments. Clinostats are still used today. By slowly rotating a plant vertically and horizontally, the clinostat prevents the plant from fixing on gravity. It then grows almost as if there were no gravity at all. Clinostats have fascinated amateur scientists for generations. Don Graham, my grandfather and an amateur scientist extraordinaire, published his explorations into geotropism in these pages 26 years ago [see pages 72–76].

Oddly, professional biologists have paid little attention to what is perhaps the most interesting region to investigate—between 0 and 1 g (the acceleration caused by the earth's gravity, equal to an increase in speed every second of 9.8 meters per second). This is good news for the amateur, for a little dedication could reward you with original discoveries.

The rotating platform of the device described here is a bicycle wheel. By properly choosing the wheel's rotation speed and placing the seeds at different distances from the pivot, you can germinate seeds at any effective gravity. Observe the thresholds at which plants first respond to gravity and see how seeds would grow on Mars (about 0.4 g). Tumbling the wheel is also necessary. The earth's gravity then averages to zero, so that the seeds consistently experience only the centrifugal acceleration from the spinning.

Besides the wheel, you will also need to scavenge parts of the bicycle frame—in particular, the front mounting forks and its hollow shaft that slides up the head tube (the part through which the handlebar stem goes). I bought them all for $15 from a bike rental shop, which kept them around as spares.

Pieces of wood support the frame and wheel. Cut a hole through the upper ends of two slats of 1-inch-thick pine shelving; attach the lower ends

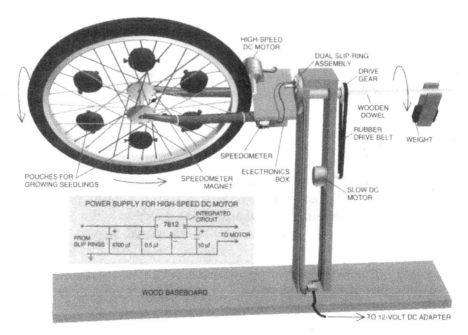

POWER SUPPLY FOR HIGH-SPEED DC MOTOR

Figure 10.4 A bicycle wheel set tumbling and spinning can convince seeds, in pouches wedged between spokes, that they are sprouting on board a spaceship. When constructing the circuit for the high-speed motor's power supply, be sure to mount the 4,700-microfarad capacitor at least 5 centimeters from the type 7812 integrated-circuit chip.

to a pine baseboard so that the slats stand upright *[see Figure 10.4]*. Thread the mounting-fork shaft through the holes. For a drive gear, try the plumbing department of a hardware store. You can make an excellent one by cutting 2 inches off the end of a large-diameter plastic or rubber pipe. Shore up the inside of the circular ring with a wooden plug. Cut a hole in the center of the plug, then thread the assembly over the shaft and epoxy it in place.

Next, thread a dowel through the drive gear and into the shaft, then attach a weight, such as those scuba divers use on their belts, to the dowel. Bolt a 3-by-7-inch wood slat to the end of the dowel. Temporarily tie the weight to the slat. Position the wheel so that the bend of the mounting forks points upward. Slide the dowel into the shaft tube and adjust the weight's position, both horizontally and vertically, until the wheel is balanced against its closest support. Epoxy the dowel in place and secure the weight with bolts.

To negate the effects of gravity, the bicycle wheel must complete a tumble about once each minute. I used a slow motor (0.28 revolution per second, or rps) and connected the motor shaft to the drive gear with a flat rubber belt (check the power-tools section of your local hardware store). So attached, the drive gear turned at the right speed. If your motor spins at a rate much different from 0.28 rps, then you will have to fiddle with the size of your drive gear or add another gear to the motor shaft to produce the correct ratio.

A second, faster motor spins the bicycle wheel to create the artificial gravity. I used a 12-volt DC, which I bought surplus and which spun at 192 rps. Make sure the tire's tread is smooth—tires from road bikes work well. The radius of the motor's shaft is about 1 millimeter. To figure out at what frequency to spin your wheel, *consult the box on this page.*

In my clinostat the radius of the wheel was 30 centimeters; at 192 rps, my motor spun the wheel at 0.64 rps. You might be able to increase your motor's effectiveness by wrapping a few layers of cloth tape around the spool; the extra material will easily boost the rotation frequency to 1.5 rps. Power comes from a DC adapter and feeds the DC motor through two slip rings. The circuit shown in the schematic *on the opposite page* regulates the power.

Commercial bicycle speedometers let you easily monitor the acceleration. Get the kind that uses a small magnet placed on one of the spokes.

Calculating G Forces

A seed on the spinning wheel experiences a centrifugal acceleration of $(2\pi f)^2 r$, where f is the rotation frequency, and r is the distance from the seedlings to the pivot. The formula to determine at what frequency to spin your bicycle wheel is $f = \frac{1}{2}\sqrt{(a/r)}$, where a is the acceleration. For example, to produce 9.8 meters per second per second (1 g) at the rim of a wheel 0.3 meter in radius requires the wheel to spin 0.91 revolution per second (rps).

The frequency at which the Radio Shack motor will drive the wheel, f_w, is $f_d (r_m/r_w)$, where f_d is the frequency of the drive motor (192 rps), r_m is the radius of the motor shaft (about 1 millimeter), and r_w is the radius of your wheel.

To find the acceleration at any distance from the wheel's center, use the formula

$$a = 19.98 \frac{V_{mph}^2}{r_w} \frac{r}{r_w} = 7.72 \frac{V_{kph}^2}{r_w} \frac{r}{r_w}$$

where r and r_w are expressed in centimeters, and v is the speedometer reading, expressed in either miles per hour or kilometers per hour.

Mounted near the axis, the magnet can measure the speed to the nearest 0.1 mile per hour or (better yet) 0.1 kilometer per hour.

So little work has been done in this area that you can grow just about anything and find something new. I've been focusing on corn. I let the seeds germinate for several days and then measured the total length of the sprouts and their "angularity"—the sum of the bend angles along the stock. At 1 g, the plants grow very straight; near 0 g, they become quite crooked. Growing seedlings at a number of locations along the radius of the wheel enables you to see the effects of gravity "turn on" inside the plant.

Put five seeds into a small handful of potting soil and place them inside the leg of an old nylon stocking. Cut the ends around the soil, then tie them off with a bit of twine to create a small pouch. These lightweight packets hold the seeds in place and make them easy to water. To get good statistics, you will need to sprout roughly 30 seedlings for each acceleration, so you should place six of these bundles at the same distance from the center. Arrange them symmetrically between the spokes of the wheel to keep the wheel balanced. You can make more bundles and insert them at different distances, so that you can experiment at different accelerations simultaneously.

Remove the seeds after they have germinated for three to seven days (or until they begin to poke out of the pouches). Cut the seedlings at their bends, then lay all the pieces end to end so that all the bends are in the same direction. The angularity is the angle between the first piece and the last piece. To measure the length of a seedling, lay a string along the piece. For each acceleration, divide the angularities by the seedling lengths and average the results. Plot this average versus acceleration, and you'll see how the sensors in your plants respond to different gravitational fields. You might also consider replanting the seedlings in normal conditions to see how they subsequently do—or try to figure out the best food plant that would grow on Mars.

11 EXPLORING GROWTH INHIBITORS

by C. L. Stong, April 1962

Here's a wonderful experiment—a truly great introduction to elementary botany. Anyone can carry out this and similar projects for almost no cost at home. All it takes is a few seeds, some tap water and a little persistence. And yet this simple experiment can open up the vast world of botanical chemistry. After all, plants can't brush away pests or run from predators, and so all species have become masters at chemical warfare. The plant kingdom defends itself with a staggering arsenal of herbicides, insecticides, and potent poisons to keep competing flora and hungry fauna at bay.

Since a herbicide must be able to get into the roots of leafy interlopers, such a toxin must be soluble in water. As a result, these chemicals are easy to retrieve by leaching in water, and that makes them ideal for home experiments.

I hope you'll experiment with many different kinds of plants. And when you think you're up to the task, see if you can isolate the chemical using electrophoresis [see Chapter 2]. Ed.

In their preoccupation with substances that encourage the growth of plants, horticulturists may overlook the fascinating experiments that can be conducted with compounds that plants manufacture to inhibit growth. These potent substances help to preserve the species that manufacture them and to regulate the density of plant populations. By producing a compound that discourages the encroachment of its neighbors a plant can provide living space for itself. Similarly, by elaborating a substance that inhibits the germination of its seeds when conditions are unfavorable the plant can ensure a good start in life for its offspring.

Substances that inhibit germination appear to be almost as various as the plants that make them, and their chemistry is equally diverse. Some of them are among the oldest and best-known drugs, stimulants and poisons in the armamentarium of medicine. The alkaloid strychnine is one. Penicillin, an unsaturated lactone, is another. Many of the germination inhibitors are also insecticides. For the most part they are cyanogens, organic acids, unsaturated lactones, aldehydes, alkaloids, and the essential oils. It is reported that even the slightest smear of oil from lemon peel, for example, will prevent the germination of wheat in a dish of otherwise fertile soil.

Inhibitors tend to concentrate in parts of a plant according to function. In the case of leafy vegetables, such as cabbage and lettuce, they are found in the leaf coat. They are concentrated in the leaf sap of spinach, in the bulb of the onion and garlic, and in the root of the carrot and the horseradish. In apples and pears they are stored in the pulp of the fruit; in tomatoes they are stored in the juice. Their function in fruits is to delay germination until after the fruit has fallen and decomposed into soil nutrient. The inhibitors are then leached away by rain in preparation for the new crop.

Some inhibitors are built into the seed, and not many of these have been investigated. Last fall Michael Zimler, a high school student in Roslyn Heights, New York, was casting about for a science-fair project and hit on the idea of setting up an experiment to learn if the seed of Merion bluegrass, the popular lawn cover, contains an inhibitor and, if so, how effective it is against the germination of other plants.

"I started out," Zimler writes, "on the assumption that the grass seed contains an inhibitor that could be extracted by water in sufficient concentration to be detected. The apparatus used in the experiments was assembled for the most part from materials found around the house: assorted glasses, bottles, jars, Saran Wrap, and toy balloons. The specimens exposed to the inhibitor included the seeds of radish, lima bean, green pea, cucumber, corn, morning glory, sunflower, zinnia, and gourd. Tests were also run on yeast and bread mold. Packets of fresh seeds were bought from a local store that deals in garden supplies, and the yeast, in dry form, came from the corner grocery. Three germinating media were used: white blotting paper, washed sand and a mixture of washed sand and peat moss.

"To make the extraction I put a half-pound of grass seed in a half-gallon jar, added a quart of tap water and let the mixture stand overnight. I stirred it occasionally before bedtime and again in the morning. At the end of 12 hours the liquor was filtered through a square of nylon mesh cut from an old stocking that had been washed with soap and thoroughly rinsed.

"Enough tap water was added to the filtered liquor to make up two quarts. This was poured into smaller jars, which were wrapped with aluminum foil to keep out the light and stored at room temperature. Within a few days the extract spoiled, turned cloudy and developed an offensive odor. I made up another batch the same way but stored it in the refrigerator at approximately 42 degrees Fahrenheit. This suppressed the growth of microorganisms, and the extract remained clear throughout the period of the experiments.

"My first attempt to germinate seeds also failed. Several conical dessert glasses were lined with white blotting paper. The paper in half the glasses was saturated with the extract; the paper in the other glasses, which were to serve as controls, with tap water. The dry seeds from the packets were inserted between the blotting paper and the glass so that sprouts could be observed without disturbing them. The number of seeds planted in each glass varied from 12 to 50, depending on the variety and size. The arrangement seemed sensible, particularly because it would be easy to keep the paper moist. It turned out, however, that only a small area of each seed made contact with the paper and the seeds did not get enough moisture to sprout.

"The next batches were planted in sand. The glasses were cleaned, dried and nearly filled with dry sand. The seeds were soaked overnight, the controls in tap water and the test specimens in extract, and embedded lightly in the sand. Thereafter the sand was kept moistened with either water or extract as appropriate and maintained at room temperature. Within a week a high percentage of all the controls had germinated with the exception of the lima beans.

"The presence of an inhibitor was strikingly apparent, particularly in the cases of cucumber, green pea and radish. Within a week 58 percent of these seeds sprouted in the control plantings but none in the sand to which grass extract had been added. Plants showing maximum resistance to the inhibitor were sunflower, corn, morning glory, and zinnia, in that order. The controls, in the case of these four plants, also exhibited more vigor than plants that were susceptible to the inhibitor. Ninety-four percent sprouted in tap water. The results are summarized in the accompanying table *[see Figure 11.1 on page 84]*, and also by graphs for morning glory, radish, and zinnia. The number of seeds in the test planting is plotted against growing time in days *[see Figure 11.2 on page 85]*.

"I performed a similar experiment to test the possible effect of the inhibitor on a fungus, which does not reproduce by seed. The most readily

NUMBER OF SEEDS	VARIETY OF SEED	MOISTENING AGENT	NUMBER OF SEEDS GERMINATED (DAYS)							
			1	2	3	4	5	6	7	8
30	CORN	WATER	0	5	17	27	28	28		
		INHIBITOR	0	0	2	9	11	11		
30	CUCUMBER	WATER	0	5	7	10	12	13	13	
		INHIBITOR	0	0	0	0	0	0	0	
11	GOURD	WATER	0	0	0	4	7	7		
		INHIBITOR	0	0	0	2	2	2		
30	GREEN PEA	WATER	0	3	5	7	7	11	11	
		INHIBITOR	0	0	0	0	0	0	0	
50	MORNING GLORY I (BLOTTING PAPER)	WATER	0	0	26	30	47	47		
		INHIBITOR	0	0	5	9	15	16		
30	MORNING GLORY II (SAND)	WATER	6	26	27	27				
		INHIBITOR	0	2	6	6				
50	RADISH	WATER	0	6	36	43	48	48		
		INHIBITOR	0	0	0	0	0	0		
15	SUNFLOWER	WATER	0	6	9	12	14	14		
		INHIBITOR	0	1	2	3	3	8		
30	ZINNIA	WATER	0	0	20	24	25	26	28	28
		INHIBITOR	0	3	4	4	4	8	9	9

Figure 11.1   Summary of effects of inhibitor on seed germination.

available fungus that can be procured in a relatively pure strain is ordinary baking yeast. Most yeasts can live and grow only in a solution that contains sugar or substances that are easily converted into sugars. Such substances are present in wheat flour.

"Two packages of active dry yeast were dissolved, one in eight ounces of warm inhibitor solution and the other, for a control, in eight ounces of warm tap water. Each yeast solution was then mixed with eight ounces of wheat flour, and the doughs were set to rise in a warm oven for one hour. The dough that was prepared with inhibitor appeared to rise more rapidly and to a somewhat greater volume than the control, but the rates were difficult to measure.

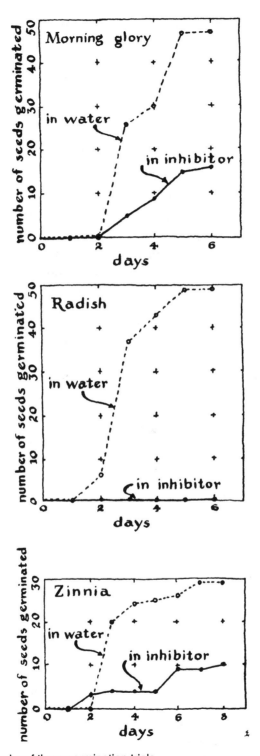

Figure 11.2 Graphs of three germination trials.

85

"Accordingly a second experiment was set up in which the influence of the inhibitor was judged by the amount of carbon dioxide liberated by fermentation. A package of yeast was divided into equal parts and each part was softened, one with 2 ounces of inhibitor liquor and the other with 2 ounces of tap water. After standing undisturbed at room temperature for four hours the yeast was further diluted so as to make 10 ounces of inhibitor solution and 10 ounces of control respectively. Four tablespoons of granulated sugar had been added previously to each of the diluting solutions as nutrient. The solutions were transferred to 12-ounce soda bottles and capped by rubber balloons from which the air had been squeezed. The capped bottles were then immersed to their necks in a pan of warm water (about 100 degrees Fahrenheit) and incubated for two hours. To maintain the temperature a small amount of cooled water was dipped from the pan occasionally and replaced by hot water. Carbon dioxide, evolved by fermentation, inflated the balloons. The volume to which the balloons expanded could be calculated approximately from measurements of their height and diameter. The calculated volume was taken as an index of the effectiveness of the inhibitor. The 'inhibited' yeast turned out to be approximately 30 percent more active than the control! Germination inhibitors, according to the literature, can affect organisms, both plant and animal, in various ways according to dosage and the nature of the organism. Caffeine, for example, will act as a stimulant, a poison, or a germination inhibitor, depending on the amount of caffeine administered, how, and to what. Perhaps the substance in Merion bluegrass acts as a stimulant for cultured yeast. On the other hand, it is possible that the inhibitor in bluegrass has no effect on the yeast; that nutrients washed from the seed account for the accelerated growth.

"I have not had time so far to check these guesses by experiment. The tests that have been described were made last year while I was a sophomore and are being continued this year on a number of molds and bacteria. Before the runs are finished I hope not only to resolve the question of whether the extract can function as a stimulant but also to identify the inhibiting substance. In several closely related plants the inhibitor has been identified as coumarin, an unsaturated lactone that is available commercially. I intend to make simultaneous runs with coumarin and Merion bluegrass inhibitor on a number of organisms and, having tabulated the results, to analyze the two solutions by paper chromatography. *[See Chapter 24.]* If a chromatographic zone of the extract migrates at the characteristic rate of coumarin and exhibits the same inhibiting properties, the extract will probably be coumarin. If no zone migrates at the rate of coumarin, the extract will be compared with other known inhibitors."

12 EXPERIMENTS WITH SUBSTANCES THAT STIMULATE PLANT GROWTH

by C. L. Stong, December 1958

When young rice plants are infected with the fungus *Gibberella fujikuroi,* they grow extraordinarily tall. In 1928 a Japanese plant physiologist discovered that an extract of the fungus produced the same effect, and shortly before World War II workers at the University of Tokyo isolated the active substance in the extract. The substance, called gibberellic acid, became the subject of intense study by plant physiologists all over the world. It has been found to exert striking influences in a wide variety of plants. When young citrus trees, for example, are treated with gibberellic acid, their stems elongate more than six times!

Gibberellic acid [one of a family of over 60 hormones called the gibberellins. Gibberellic acid is also known as gibberellin A3. Ed.] is now commercially available. It should interest amateurs because it is inexpensive, produces spectacular effects and offers an unusual opportunity for original experiments. Among those who have worked with it are Robert Lawrence and Henry Soloway, who are students in the College of Medicine of the State University of New York.

"A speck of gibberellic acid smaller than a grain of sugar," they write, "has turned our window box into an Alice-in-Wonderland jungle. The compound causes most plants to grow at record speed, flower in half the usual time, and bear fruit that will win first prize in any county fair. In view of these effects and their implications it is not surprising that gibberellic acid is now bringing many strange bedfellows together: the florist seeking to produce larger flowers, the farmer trying to double the production of his land, and the cancer researcher who hopes in some manner to find clues to the dynamics and pathology of growth.

"The amateur who experiments with gibberellic acid is likely to reap satisfying rewards because the field is still wide open. Most of the research now under way is centered on crop plants and flowers cultivated by commercial greenhouses. Amateurs can avoid duplicating these experiments by selecting less common plants. For example, to our knowledge no work is being done with molds, mushrooms, mosses, ferns, or fresh-water algae.

"The experiment which follows is designed to demonstrate certain basic reactions of plants to gibberellic acid. The approach is not limited to this compound; it will prove equally effective with any substance suspected of being a growth stimulator or inhibitor. Thus it can serve as a stepping-stone to other investigations.

"You will need a quantity of the acid, a set of identical plants, some inexpensive apparatus and, last but not least, a notebook. [Gibberellic acid is sold under the name of ProGibb 4%. It can be purchased from Agrobiologicals at *www.agrobiologicals.com*. They are a division of CPL Scientific Publishing Services Ltd., 43 Kingfisher Court, Newbury, Berks RG14 5SJ United Kingdom. Phone: 44 (0) 1635 574920. Other sources can be found online by searching for "gibberellic acid."]

A convenient plant for the experiment is the common garden pea, although many other plants are equally rewarding. It should be said, however, that gibberellic acid is known to have no effect on the white pine, the gladiolus, or the onion.

"You'll need at least two plants to be used in the experiment. The control plant is treated with tap water. The test specimen is treated with dilute gibberellic acid. By comparing the subsequent reaction of the test specimen with the control the experimenter draws conclusions about the effect of the acid. [Note, no one should ever draw any conclusions based on comparing just one test to one control plant. If you're to have any confidence in your results, you must compare differences between groups of plants. You should run between 10 and 30 plants in both groups, and then measure the average growth rate in each and compare these averages. Then you can not only quantify the differences between the groups, but also, by using a little statistics, you can determine how much confidence you can have in your answers. Ed.]

Your comparisons must be based on precise measurements. To start, try choosing the height of the plant, the number of its leaves, and its weight. Less obvious effects should also be selected for observation, such as the rate at which the plant consumes oxygen and the percentage of water in the plant with respect to the percentage of organic matter and inorganic ash. Once you've gained some experience, you may hit on other factors that you want to track.

"We suggest the use of 32 germinated peas as experimental plants. It is well to place 45 or 50 seeds in a germinating medium, which may consist of thoroughly moistened filter paper, a few layers of cotton cloth, or wet sawdust. The medium should be kept warm and should be covered with an inverted glass bowl to prevent evaporation. Some of the seeds may not germinate, but virtually all of those that do will mature. Do not use ungerminated seeds. This experiment seeks to measure the effects of gibberellic acid on a growing plant, not its influence on the sprouting time of seeds. The latter experiment can be equally fascinating and, incidentally, it is one that is of great interest to commercial growers.

"While the seeds are incubating, begin recording everything in a notebook. Every detail, however self-evident, should be entered along with the entry date. Reserve the first few pages for a running summary. Include the date on which the acid was received (for information in case the compound should deteriorate with time), the date on which the peas were set for germination, the temperature, the date of planting, when the acid was first administered, the date of the second treatment, and so on. It is in this information that explanations will ultimately be found of how the acid does its work.

"When the peas have germinated, 32 paper drinking cups are filled to within a quarter-inch of the rim with sifted topsoil moistened just enough to form a fragile lump when a pinch of it is squeezed. One sprouted pea is then planted in each cup with the tip of the shoot pointing up and flush with the surface of the soil. The cups are arranged in groups of four and labeled according to the concentration of gibberellic acid the group is to receive. One group of four cups is reserved as a control and receives only tap water.

"The gibberellic acid comes mixed with an inert filler. If the experiment were ideally controlled, another group of cups would be reserved for treatment with the filler alone. For this experiment, however, we assume that the manufacturer's filler is really 'inert.'

"You'll need seven small bottles with a capacity of 3 or 4 fluid ounces for storing dilutions of the acid. These may be purchased for a few cents at most drugstores. The kind which has a scale of cubic centimeters molded in the glass is convenient. If bottles with such scales are not available, the experimenter must either make or buy a graduate.

"The dilutions are prepared by first dissolving 1 teaspoon (2.5 grams) of

ProGibb 4% in 50 cubic centimeters of tap water. Each teaspoon contains 5 milligrams of gibberellic acid; hence each cubic centimeter of this solution will contain 1 mg. of acid. The second dilution is prepared by pipetting 5 c.c. of the first dilution into another container and adding enough tap water to make 50 c.c. Each cubic centimeter of this solution contains 0.01 mg. of acid, and each dose of 10 c.c. contains 0.1mg. A dilution containing 0.01 mg. per 10 c.c. is made by pipetting 5 c.c. of the second dilution into a third container and adding water to make 50 c.c. The process is continued until seven dilutions are prepared so that 10 c.c. of each contains, respectively, 1 milligram, 0.1 mg, 0.01 mg., 0.01 mg., 0.001 mg., 0.0001 mg., 0.00001 mg. and 0.000001 mg. of the acid. Label each bottle to indicate the dilution it holds. Fresh dilutions must be prepared for each treatment, because the acid gradually loses its activity in solution. [It doesn't take much. You should find activity at the few parts per million level. Commercial applications often spread just a few grams of the acid on each acre of farmland. Ed.]

"To eliminate the influence of variations in environment during the experiment, the growing plants should be placed in a dark room in which the temperature does not vary appreciably from 70 degrees Fahrenheit, and should be exposed daily for 11 hours to a fluorescent lamp of at least 40 watts placed lengthwise above the cups at a height of two feet. [Alternatively, you can use fluorescent Grow Lamps. These were designed with horticulturists in mind. They reproduce the sun's spectrum better than ordinary fluorescent bulbs. You'll find them at any well-stocked nursery. Ed.] Each experimental plant receives 10 c.c. of the appropriate acid dilution, as indicated by its label, every 48 hours. No water or other solution should be administered to them. The four control plants receive 10 c.c. of tap water at the same time. The acid should always be administered in a uniform manner. Ideally it is applied as a spray by means of an atomizer, which assures that all the exposed parts of the plant receive the solution. If desired, however, the dilutions may be poured on the soil near the base of the plant. Reaction between the acid and the soil tends to lower the activity of the acid somewhat.

"Record the height of each plant daily, beginning with zero inches on the day the experiment starts, when the shoots are flush with the top of the soil. From these data graphs may be plotted either for the individual plants or as averages of the groups. Similar tables should be constructed for the remaining indices of growth, such as the number of leaves. Weight need not be measured daily. This is a tedious operation. But the plants should be weighed individually at the conclusion of the experiment. [These days, I place all my data in a computer spreadsheet and periodically print it out and paste the table into the notebook. These programs will save you a lot of time and ensure accuracy when you want to do mathematical calculations, like finding averages and standard deviations. They will also plot the data directly for you! Ed.]

"Gibberellic acid appears to increase the rate of the metabolism of plants. It is possible to investigate metabolism by measuring the rate at which the plants consume oxygen. *[The devices described in Chapter 3 can be adapted to measure plant metabolism, as well that of animals. Ed.]* The plant is enclosed in a vessel to which air can be admitted in accurately measured amounts. The vessel contains a small quantity of calcium chloride (soda lime) [sodium hydroxide works as well. Ed.] to absorb carbon dioxide liberated during respiration. Plants take up oxygen from the air and give off carbon dioxide only during intervals when photosynthesis is not taking place. In determining oxygen consumption, therefore, all measurements must be made when the plant is in the dark. The enclosing vessel may be opaque or, if it is made of glass, should be kept in a dark room.

"An adequate vessel can be made by upending a two-gallon tin can on a plate of flat metal. Hose connections can be introduced through the metal bottom. After the plant and calcium chloride are in place, the assembly is made airtight by applying a ring of wax (made of equal parts of beeswax and rosin) between the can and plate as shown in *Figure 12.1 [page 92]*. The wax, which should be applied smoking hot with an eyedropper or a small measurement and reused. One hose connects the vessel with a U-shaped length of glass tubing which serves, when partially filled with colored water, as a manometer for measuring the difference in pressure between the closed vessel and the room. A second hose terminates in a calibrated hypodermic syringe by which air is admitted to the vessel in measured amounts. The third hose serves as a vent for the vessel and is normally kept closed by means of a pinchcock. The potted plant may be supported on a sheet of stiff wire screening placed over the dish of calcium chloride. Unless the seams of the can are airtight they should be coated with wax. A coating of Vaseline will seal the piston of the syringe to the walls of the cylinder. [Note: An old pressure cooker makes an ideal vessel here. It's already airtight and easy to machine. Ed.]

"To make a measurement, a fresh supply of calcium chloride is placed in the dish, the plant is set on top of the screen and covered by sealing the tin can in place. The piston of the hypodermic syringe is placed at the 10 c.c. graduation. The pinchcock is opened until the columns of colored water in the arms of the manometer stand at the same height, which indicates that the air pressure in the vessel and the room are equalized. The

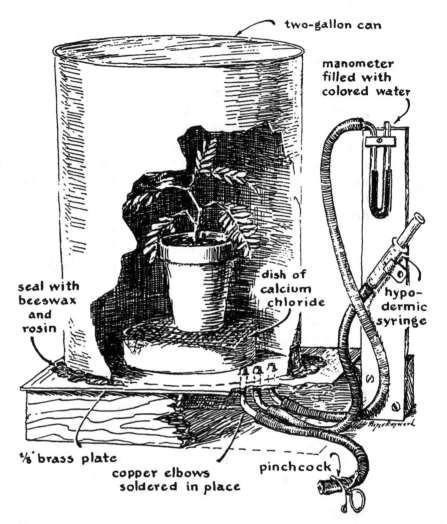

Figure 12.1 A homemade apparatus to measure the oxygen consumption of plants.

pinchcock is then closed and the time is recorded. After an interval which depends on the size of the plant and the rate of its respiration, the plant will take up enough oxygen from the air to cause an appreciable drop in the pressure indicated by the manometer. The piston of the syringe is then pushed in until the columns of water in the arms of the manometer again stand at the same level. The time is then recorded along with the volume of air admitted to the vessel from the syringe. The volume of air required to equalize the pressure is proportional to the oxygen consumed during the interval, and is an index of the metabolic rate of the plant. The rate of oxygen consumption is computed by dividing the volume of air admitted from the syringe into the vessel by the elapsed time in minutes. Accuracy can be improved by making three successive tests of the same plant and averaging the consumption rate of the three runs. The experiment may extend over several weeks, during which atmospheric conditions as well as the temperature change; for this reason tests made on different days may not be comparable. It is therefore necessary to add a correction which adjusts the figure to the value it would have if the test were made at 'standard' temperature (20 degrees Centigrade) and barometric pressure (760 millimeters of mercury). To make this adjustment, the average amount of air admitted from the syringe to the vessel is multiplied by the barometric pressure (measured in millimeters of mercury) and divided by 760. This quotient is then multiplied by 273 and divided by 273 plus the temperature of the room in degrees Centigrade. [Note, the adjustment for temperature must be made on the scale of absolute temperature scale, that is, the Kelvin scale. Twenty degrees Centigrade is 293 Kelvin. Ed.] The result is entered in the notebook for the plant and dated.

"The oxygen consumption of plants varies as a logarithmic function of their surface area. Hence, for strictest accuracy, only plants of identical surface area can be compared. It is not easy to determine the surface area of a plant, but by making the assumption that all plants of equal height have comparable surface areas one can make comparisons which are interesting and useful even though approximate. *[See sidebar, page 94, on how to measure the area of a plant. Ed.]*

We have assumed in this experiment that all the plants that are three inches high have the same surface area. The oxygen consumption of the group receiving the highest concentration of gibberellic acid is measured when these plants reach a height of three inches. A day or so later the next group will have reached the same height and can be similarly measured. The test is repeated as each of the more slowly growing groups reaches the height arbitrarily selected. The rate of consumption is plotted (on the vertical coordinate of a graph) against the logarithm of the dose (on the horizontal axis). The logarithm of one milligram per day equals zero, the log of 0.1 mg. per day equals −1, of 0.01 mg. per day equals −2, and so on.

"At the conclusion of the growth period, when, say, the most slowly growing group reaches a height of three inches, all the plants are carefully removed from the paper cups and gently agitated in a large pan of water until the soil adhering to the roots sinks to the bottom of the pan. Care should be taken not to tear away any of the roots. The plants are then rinsed, blotted dry and promptly weighed. Any delay may introduce error

The text claims that it's difficult to find the surface area of a plant. That was true back in the days before computers. But once again, modern technology comes to the rescue! I stand an image scanner on its side and prop it up above the table using a few books. Then I position the plant to be measured so that its pot is just below the scanner. When the leaves are gently pressed between the cover and the glass the plant's image can be scanned directly into my computer. The plant's surface area is related to the total number of pixels it occupies.

The plant's area can now be obtained in several ways. You could print the image on graph paper and count the covered squares. Or you could print the image, cut out the outline and weigh the remaining paper. If you happen to have a sensitive balance you could measure the weight of the toner on the printed page.

However, I use PhotoShop to find the pixel count. First, for reference, I always attach a gray paper disk of known area next to the plant before the scan. Once the scan is complete I remove any structures that I'm not interested in. For example, if I want to find the area of the leaves only, I delete the stems from the image. Next, I fill the image with black so all the pixels are uniformly dark, and I fill the reference disk so that it is uniformly gray. Then I go to the Image menu and select Adjust > Levels. PhotoShop puts up a dialog box. If the disk is all one color and the plant another, you'll see two vertical lines, indicating how many pixels are set to each hue. Then just lay a flexible plastic ruler over your screen and measure the height of each line. The area of the plant is the area of the disk times the ratio of the heights of these two lines (plant over disk).

Some systems crowd more pixels in the horizontal direction and the vertical. So it's a good idea to flip the image 90 degrees and repeat the measurement. If you get a different answer, then average the two measurements and use the average as the plant's area. Ed.

because the plants will lose water through evaporation. The weight of each plant is recorded. Then the leaves and roots are cut from the stem and weighed separately. The combined weight of the parts should nearly equal the total weight of the plant. Evaporation may account for a slight difference, but any substantial disagreement may indicate an error in procedure. Graphs of the weight are then drawn, in which the total weight is plotted against the logarithm of the dose, as in the case of oxygen consumption. Similar graphs should be made showing the percentage of weight as a function of dose represented by the leaves, stems and roots respectively.

"Most plants consist mainly of water, and it is important to discover if the

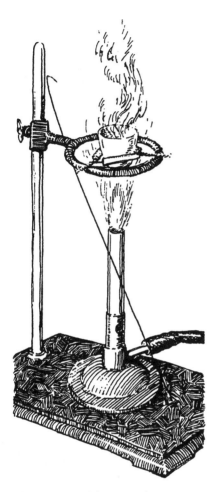

Figure 12.2 A homemade apparatus to measure the oxygen consumption of plants.

growth induced by gibberellic acid represents an increase in the solids or merely an increase in water. This can be determined by thoroughly drying the plants and comparing the dry weight with the total weight previously recorded. Separate tests should be made of the leaves, roots and stems to disclose the effects of the acid on the several parts of the plant. Plants may be dried in about three days by placing them on top of a radiator or under a 100-watt incandescent lamp. To find the percentage of water, multiply the dry weight by 100, divide by the wet weight and then subtract the quotient from 100.

"How does gibberellic acid affect the rate at which plants take up inorganic substances? This can be investigated by burning the dried remains and weighing the ash. In making this test it is again interesting to measure the leaves, stems and roots separately. An accurately weighed sample is placed in a crucible supported in the flame of a Bunsen burner *[see Figure 12.2]*. Set the crucible somewhat obliquely in an asbestos triangle to allow for the expansion of the heated parts, and close it with a loosely fitted cover. The crucible is then maintained at a red heat until the contents turn to a white, powdery ash. When cool, the ash is transferred to a balance and weighed. Multiply the weight of the ash by 100 and divide by the weight of the dried material placed in the crucible. This gives the percentage of inorganic ash.

"From the accumulated data it is now possible to answer the following questions: Does gibberellic acid increase the rate at which the experimental plant stores energy or merely cause it to absorb an abnormal amount of water? How does gibberellic acid affect the plant's consumption of oxygen? What further experiments do the results suggest?"

13 THE EFFECTS OF ULTRASONICS ON PLANT DEVELOPMENT

by C. L. Stong, August 1966

In recent years several experimenters have observed that ultrasonic vibrations in air stimulate the growth of plants. The effect appears at frequencies above 20,000 vibrations per second and levels off above 50,000 vibrations. The possibility that plants might respond to sound waves was recognized more than a century ago. Charles Darwin attempted without success to stimulate the "sensitive plant" (*Mimosa pudica*) with sound waves generated by the bassoon and other musical instruments. Similarly negative results were subsequently reported before 1900 by many other experimenters, including the eminent German plant physiologist Wilhelm Pfeffer.

In the light of these reports interest in such experiments largely disappeared until the advent of electronic apparatus for generating ultrasonic vibrations. Last year Evalyn Horowitz, then a high school student in Bergenfield, New Jersey, won a prize at a science fair for a project based on the exposure of radish seedlings to vibrations of 50,000 cycles per second at an acoustic energy equivalent to about one watt. She demonstrated that the rate of growth is almost doubled during the three weeks following germination and that accelerated growth is observed even when the plants are kept in darkness.

The ultrasonic generator used by Miss Horowitz consisted of an electronic oscillator connected to a small loudspeaker of the kind used for the "tweeter" in high-fidelity phonographs. "The idea of doing the experiment," Miss Horowitz writes, "came to me when one of my teachers explained that the growth of plants can be influenced by light of a selected color. Somehow the words 'light' and 'sound' became linked in my mind; I wondered if sound of a certain pitch might not have a comparable effect on plant growth. The literature was not encouraging. Most investigators who had subjected plants to vibrations in the audible range reported negative results. A bulletin issued by the U.S. Department of Agriculture, however, stated that 'the effects [of sound] on flowering, growth and yields have not yet been evaluated with respect to time of treatment and intensity of treatment.' This implied that the question was not settled. So I decided to set up an apparatus for exposing potted plants to ultrasonic vibrations. I chose radishes for the experiment primarily because they are hardy and the seeds were available.

"Two groups of plants were grown for each experiment. One was exposed to ultrasonic energy and the other was insulated as a control. The apparatus used in the first series of experiments consisted of two cabinets with a volume of 1 cubic foot each. They were lighted by 7.5-watt incandescent lamps. The experimental cabinet also housed a tweeter, which is a small loudspeaker powered by an audio generator of the kind used by electronic-service technicians. The audio generator developed a maximum frequency of 50 kilocycles at about 1 watt. I bought these components, together with timing switches, from a dealer in amateur radio supplies.

"The cabinets were constructed of acoustic ceiling tile 2 inches thick. I obtained the tile from a lumberyard. The material was cut to size and assembled with wood screws. A small hole in the top of each cabinet admitted a thermometer for monitoring the temperature of the air. The assemblies were closed with snugly fitting, removable covers of the same material *[see Figure 13.1 on page 98]*.

"I later learned that the sound would have been more effectively confined if I had made the cabinets by nesting two plywood boxes with an inch of space between them and filling the space with sand. My construction worked, however, perhaps because during operation the cabinets were separated by about 10 feet. At this distance little ultrasonic energy from the experimental cabinet penetrated the control cabinet.

"Enough garden soil was procured for eight clay pots of two-inch diameter. The soil was thoroughly mixed in a cardboard box, tested for chemical composition, fertilized, moistened, and packed lightly into the pots. Four radish seeds were planted in each pot at a depth of about an eighth of an inch. All the pots were watered equally and simultaneously.

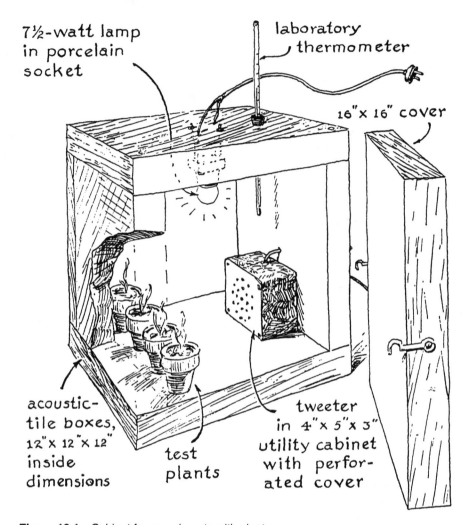

7½-watt lamp
in porcelain
socket

laboratory
thermometer

16"x 16" cover

acoustic-
tile boxes,
12"x 12"x 12"
inside
dimensions

test
plants

tweeter
in 4"x 5"x 3"
utility cabinet
with perfor-
ated cover

Figure 13.1 Cabinet for experiments with plants.

The time switches were set to turn on the lights and the audio generator at 8:30 A.M. daily and turn them off 12 hours later.

"Records were made of the dates of planting and sprouting. When the seeds germinated, each plant was identified by a numbered label affixed to the edge of the pot. Thereafter on Mondays, Wednesdays, and Fridays notes were made of the general appearance of each plant, including height, leaf growth, and coloring. The plants were measured in two ways: by straightening the stems and placing them against a ruler and by estimating the height against a background of metric graph paper. I am not satisfied with either method. The first entails handling the plants, with the risk of damage, and the second is subject to error. I did not succeed, however, in devising a better measuring scheme.

"The first experiment was stopped after 28 days, and I then plotted graphs of the relative growth of the experimental plants and the controls *[see Figure 13.2].* Plants that were treated with ultrasonic vibrations grew an average of 87 percent taller than the controls. In general appearance the controls were small and sturdy, with thick stems and bright green leaves. In contrast, the experimental plants were comparatively spindly. They were tall and shaky, with thin stems and dark green leaves.

"This experiment demonstrates that exposure to ultrasonic vibration affects the growth of radishes, but it does not explain why. It has been suggested that the vibrations 'awaken' the plants, that is, that they first stimulate the metabolic processes as light becomes available for photosynthesis and that later they accelerate the biochemical reactions. If this were the case, the plants should produce more and larger leaves and thicker stems; in general they should become large counterparts of the controls. Instead they merely grow taller.

"The theory has also been advanced that the vibrations raise the temperature of the soil, thereby increasing the activity of soil microorganisms and providing the experimental plants with an abnormal amount of nutrients. A careful check of soil temperature by thermometers that could be read to within half a degree failed to show a significant difference in soil

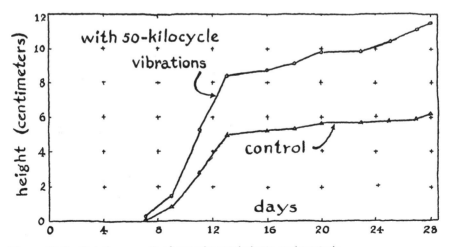

Figure 13.2 Relative growth of experimental plants and controls.

temperature between the two cabinets. Increased microbe activity should change the acidity of the soil. A careful check of soil acidity at the end of the experiment indicated a pH of between 3.5 and 4 in all pots.

"The plant hormones known as auxins can encourage elongated growth. It occurred to me that the most common auxin, 3-indoleacetic acid (IAA), might be involved. Perhaps the vibrations either increased its formation and activity or decreased the effectiveness of its inhibitors. As a rough check on this hypothesis I made up four solutions of the chemical: two test tubes containing five grams of a 0.1 percent solution of IAA (the approximate concentration found in plants) in ether, and another two test tubes containing five grams of a 1 percent solution of IAA in ether. One tube of each concentration was placed in the control and experimental cabinets respectively. The experimental solutions were exposed to 50-kilocycle vibrations 12 hours a day for 10 days. The 0.1 percent experimental solution lost 0.08 gram more IAA than its control and the 1 percent solution lost 0.17 gram more than its control. These losses do not necessarily imply increased chemical activity as a result of the ultrasonic irradiation, but they indicate that some phenomenon other than evaporation is at work.

"In another experiment I compared the growth of two groups of radish seedlings: one treated with IAA and exposed to ultrasonic vibrations, the other not treated with the hormone but exposed to the vibrations. Two other groups, one treated with the hormone and one untreated, were used as controls. These plants were potted in troughs that required more room than was available in the original cabinets. Larger cabinets were built [see Figure 13.3]. These were similar to the ones made first but were lighted by 20-watt fluorescent lamps. The lamps overheated the cabinets. This problem was solved by placing the ballast coils of the lamps outside the cabinets and by keeping a tray of ice inside as required. A more elegant cooling system could have been set up by installing a heat exchanger in the cabinets, but my scheme worked nicely even though it required close attention.

"One group of control plants and one group of experimental plants were watered regularly, as in the first experiment. The other group of controls and the second group of experimental plants were watered with a 0.1 percent solution of IAA. All other conditions were identical with those of the first experiment. The IAA-treated control plants showed 91.5 percent of the growth of the untreated controls. The IAA-treated experimental plants grew 140 percent higher than the untreated controls, and the untreated experimental plants grew 150 percent higher than the untreated controls.

"Although the hormone-treated plants in both cabinets grew somewhat less than their untreated counterparts, the treated experimental

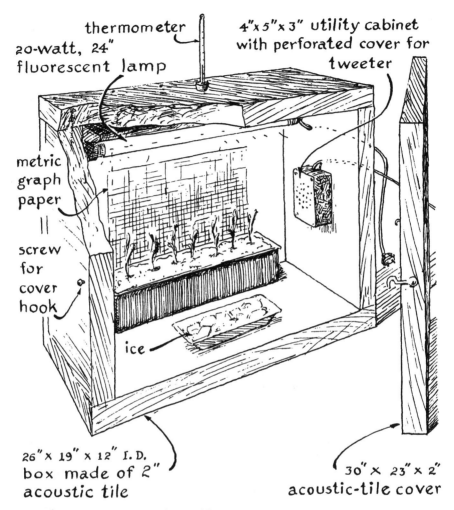

thermometer

20-watt, 24" fluorescent lamp

4"x 5"x 3" utility cabinet with perforated cover for tweeter

metric graph paper

screw for cover hook

ice

26" x 19" x 12" I.D. box made of 2" acoustic tile

30" x 23" x 2" acoustic-tile cover

Figure 13.3   Second design for a cabinet.

plants exhibited more growth with respect to the untreated experimental plants than the treated controls did in relation to their untreated counterparts. This result does not necessarily implicate IAA in the phenomenon of growth acceleration, although in my opinion some relation is implied. The lower comparative growth rate of the treated plants may be explained by the fact that auxin can cause a plant to spend more energy 'burning' food than using it.

"As a final experiment in the series, I compared the relative growth of a group of normally lighted controls with a comparable group of experimental plants exposed to ultrasonic vibrations but deprived of light. I was

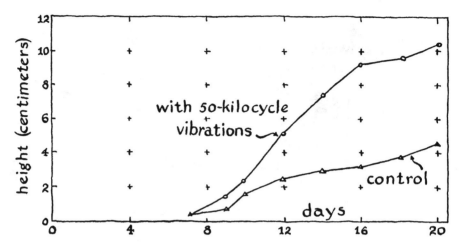

Figure 13.4   Results of further experiments.

curious to learn if some or all of the energy necessary to support growth could be provided by ultrasonic vibrations. Plants normally grown in darkness are tall, thin, and devoid of the green pigment chlorophyll; they die of starvation soon after sprouting. Unfortunately this experiment was undertaken just prior to the science fair, when my time was limited. Even so, it produced at least one interesting result.

"The control cabinet was lighted by a single 7.5-watt incandescent lamp. The experimental plants were irradiated with ultrasonic vibrations in darkness. Both cabinets were operated 12 hours a day for 20 days. Four plants were grown in each cabinet. The average growth of the two groups is shown in the accompanying illustration *[see Figure 13.4]*. The control plants were small but sturdy; they had thick stems and the leaves were bright green. The experimental plants, which grew 134 percent taller than the controls, were spindly, thin-stemmed and yellowish, but they survived."

PART 3

CELLULAR BIOLOGY

14 THE PLEASURES OF POND SCUM

by Shawn Carlson, March 1998

If extraterrestrial scientists were ever to visit our planet, they might not accord as much attention to the relatively large, complex creatures like ourselves, as we think these organisms warrant. Rather I suspect they would notice that the greatest diversity, and indeed most of the biomass, on the earth comes from its simplest residents—protozoans, fungi and algae, for example. By just about any objective measure, these organisms represent the dominant forms of life.

The world of algae is a particularly fascinating realm for amateur exploration. A single drop of pond water can harbor a breathtaking variety of microscopic species, with each algal cell constituting a complete plant. So if you want to view the essential biological underpinnings of all plant life, algae provide the best show in town.

Obviously, you will need a microscope to see the spectacle. An instrument with a magnification of about 120 times should be adequate. But the real secret to exploring this enthralling world successfully is to create small enclosures—"microponds," if you like—in which your specimens can flourish.

Canning jars filled with some water and soil make ideal environments *[see Figure 14.1 on page 106]*. To let gases in and out, cut a hole one centimeter (about half an inch) in diameter in each lid and plug the opening with a wad of sterile cotton. You will also need a source of light for these diminutive plants. Because direct sunlight can rapidly warm your tiny ponds to temperatures that are lethal to algae, place them in a window with a northern exposure. Or set a full-spectrum bulb on a timer to provide a more controlled source of illumination.

You must sterilize the soil, water, jar, and lid before introducing any algae; otherwise stray bacteria could quickly take over. A special apparatus called an autoclave kills bacteria with heat under enough pressure to keep water from boiling. Scientists at a local university or commercial laboratory would probably be delighted to sterilize some materials for your home research using an autoclave. But an ordinary pressure cooker will also work.

You should first assemble everything. Add about four centimeters of dirt or mud and distilled water fortified with commercial plant food according to the manufacturer's instructions. Fill the jar to within two centimeters of the top, but do not screw the lids down yet. Place the prepared jars inside a plastic basin that is filled with two centimeters of water to

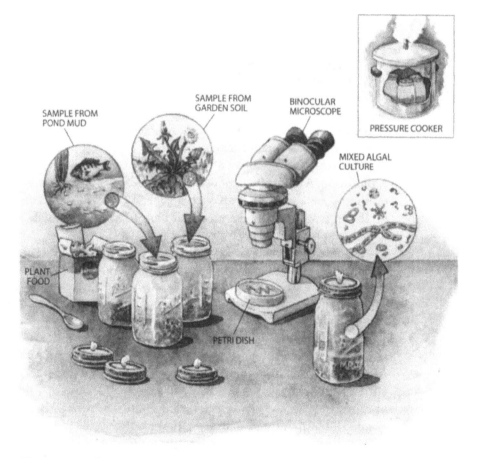

Figure 14.1 Sterilized canning jars make ideal incubation chambers for growing algae taken from a garden or pond.

prevent the base of the glass from getting too hot and breaking. Heat the jars for 20 minutes at 120 degrees Celsius (about 250 Fahrenheit).

If you don't have access to a pressure cooker, you can try vigorously boiling the water, jar and lid for at least 20 minutes in a covered pot and baking the soil in your oven at 180 degrees C for one hour. This method requires that you assemble the sterilized parts afterward, a procedure that risks contaminating your cultures with bacteria. If you do use this approach, let your containers sit for at least 10 days to make sure that your ponds are truly sterile; contaminating bacteria will turn the water cloudy as they multiply.

Otherwise, once the microponds have cooled completely, you should inoculate them with algae as soon as possible. A scraping from a piece of seaweed, a smidgen of pond mud, a pinch of garden soil, even a rubbing from the inside of a friend's aquarium are all great sources. Add such samples to a few jars and watch these aquatic gardens grow.

Your first cultures will probably contain a jungle of different single-celled plants. For scientific work, you will need to isolate individual species. You can do so by separating a small group of cells and implanting them in sterile agar, a gelatinous growth medium, which, incidentally, is itself made from algae. (A source of agar is listed at the end of this chapter.)

This process is not as hard as it sounds. You begin by filling a set of sterilized petri dishes with agar to which you've added a small amount of plant food. Add the food just before setting the agar aside to gel. Next, use a sterile probe to take a tiny piece of the algae from one micropond. Swab the probe rapidly over the surface of the agar in a widely spaced zigzag pattern. Cells transferred onto the nutritious surface will take hold and grow in a few days into a splotchy garden of isolated groups plainly visible to the naked eye. Because cells will tend to shed in clumps from the original grab, most of the second-generation crop will be only slightly less diverse than their parent sample. But by taking a small amount from the center of one of these groups and then repeating the process on a second petri dish, you will create third-generation growths of even fewer species.

Some biologists recommend raising each succeeding generation by transferring a speck from a single petri dish to a sterilized culture jar, waiting for plants to multiply, and then separating them again on a fresh agar surface. But I can sometimes get results more quickly by raising multiple generations on enriched agar, without culturing in jars as an intermediary step. Try this shortcut, but don't be surprised if you find you need to use the more laborious procedure.

It usually takes between three and five generations of swabbing and growing, but eventually your microponds will each contain a single algal species. Careful examination of samples under the microscope will reveal whether your cultures are pure. Once you have isolated several strains of algae in this way, the investigations you can make are limited only by your imagination. One useful class of experiments involves seeing how different chemicals affect growth.

Your test results will be much simpler to compare if you begin each trial with samples containing roughly the same number of cells. One way to make up such samples is to dip blotting paper (sterilized with an autoclave or pressure cooker) into a dilute solution of sterile plant food and let it dry. Then soak the paper in molten beeswax for a few seconds until it is thoroughly impregnated. Remove the strips from the wax and return them to the solution of sterile plant food to harden. Transfer the waxed strips quickly to your culture jars. With luck, algal cells will grow uniformly over the treated paper. You can then cut out standardized samples using a sterile metal hole punch.

To start your experiment, set up a rack of 10 test tubes. Place a high concentration of the chemical you wish to evaluate in the first tube and then dilute each succeeding test tube by a factor of 10. That is, add nine parts of distilled water to one part of solution. (An eyedropper makes this task easier.) With 10 test tubes, you'll have a billionfold difference in concentration between the first and last mixtures.

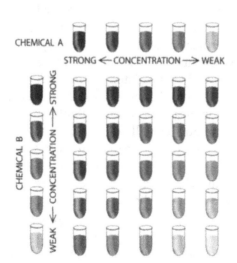

CHEMICAL A

STRONG ← CONCENTRATION → WEAK

CHEMICAL B   WEAK ← CONCENTRATION → STRONG

Figure 14.2 Variable concentrations of two substances can be prepared by mixing equal amounts from separate dilutions of each chemical.

If you want to see whether the chemical in question improves growth, add three standard hole-punched samples and a fixed quantity of each of the dilutions to sterile culture jars. If you want to test whether the chemical can kill algae, you should place three standard samples for a prescribed time in each tube. Then use sterile tweezers to retrieve the samples and wash them gently with distilled water before placing them in the sterile culture jars.

Now monitor how the treated algae grow. Once you know roughly the minimum level at which the test chemical alters the amount of algae seen in the jar after one week, you can conduct a more tightly focused experiment to determine the critical concentration more precisely. A similar procedure lets you explore how algae respond to two separate chemicals *[see Figure 14.2]*. This tactic is useful, for example, if you want to compare the effects of, say, a nitrogen-based fertilizer and a substance such as aspirin, which reduces the surface tension of water and so might make it easier for nutrients to get inside the cells.

This procedure will let you identify the optimal concentrations of two chemicals for growth of a particular species. You might also try to gauge how varying levels of heat and light affect the growth of algae, to learn their method of reproduction by microscopic examination or study their metabolism. *(See Chapter 8 for details. Ed.)* Remember, you can dive into this wondrous aquatic realm as deeply as you like.

[Powdered agar can be obtained from the Society for Amateur Scientists. Call 401-823-7800 for details. Ed.]

15 EXPERIMENTS WITH ANIMAL CELLS

by C. L. Stong, April 1966

The cells of many animal tissues can be kept alive for long periods outside the animal by means of the technique known as tissue culture. In some cases the original cells maintain themselves without dividing; in others they divide repeatedly. Essentially tissue culture consists in transferring cells from an animal to a glass vessel containing the appropriate nutrients at body temperature. One line of cells established in this way has been maintained for almost 20 years, long after the death of the animal from which they came. Such cells maintain their vitality and show no evidence of aging. Tissue culture thus offers insights into some intriguing questions. Do the cultured cells acquire a longevity they lack when they are part of an animal? Or is this longevity an intrinsic property of the cell, one that disappears when the cell functions as a member of the highly organized and complex cellular community that constitutes the intact animal?

Once experiments with tissue cultures were all but closed to amateurs. Mastery of the essential procedures required a long apprenticeship. The ingredients of the nutrient mixtures were difficult to obtain and even more difficult to compound. The procedure called for rigorous routines to prevent the infection of cultures by bacteria.

Today ready-made ways of solving such problems are at hand. Nutrient mediums can be bought inexpensively from distributors of biological supplies. Bacterial infections are controlled by antibiotics. Cultures of living cells can also be bought. With such materials Ted M. Fancolli, who attends the American River Junior College in Sacramento, California, has developed simplified methods of making tissue cultures. The methods require little more skill than growing bacteria on a plate of nutrient agar. In effect these procedures place in the amateur's hands a powerful tool for looking into such diverse matters as the structure and function of cells, the susceptibility of cells to various bacteria and viruses, the nutrition of cells and the effects of drugs and radiation on cells.

"Cell cultures," Fancolli writes, "have been grouped in three classes according to the origin of the cells and their behavior. Those established directly from animal tissue are known as primary cultures. Examples are cells from the kidneys of rhesus monkeys that are used in the production of poliomyelitis virus for both the Salk and the Sabin vaccines. Primary cultures can be established from almost any kind of tissue, but they must be prepared fresh each time they are used, necessitating a constant source of tissue.

When this article first appeared almost everyone knew about Salk and Sabin's work to eliminate polio. But since most people who are alive today weren't born when these two men made their breakthroughs, here's a little historical note that will help put this column in perspective. The first polio vaccine was developed by Johannas Salk in 1952. He grew three common strains of the virus on tissue cultures established on the kidney cells of monkeys. He fashioned his vaccine from virus that he killed by exposing it to formaldehyde. Clinical trials begun in 1954 proved so successful that the U.S. government approved the vaccine for widespread distribution almost immediately in 1955. A new, more potent version of this vaccine, one that is grown on cultivated human cells, was introduced in 1987. A second approach to protection against the polio virus was developed by Albert Sabin. His vaccine was developed in 1958 and is given orally. It also was developed from three strains of the virus grown in monkey kidney cell tissue cultures and contains live but damaged virus that colonizes the intestines, just like the dangerous strain of the virus does. Ed.

"The cells of some primary cultures can be serially subcultivated. They are then known as cell strains and will usually persist through 40 or more generations before dying out. In form and structure, cell strains do not differ significantly from primary cultures.

"For reasons that remain obscure the cells of some strains continue to reproduce indefinitely. Such cultures are known as cell lines. The oldest culture of this type, called the *L* line, was established in 1947 by Wilton R. Earle of the National Cancer Institute from tissue taken from a male mouse 100 days old. The oldest culture of malignant origin is the 'HeLa' cell line taken from a human cancer in 1952. It has since become one of the most extensively investigated cell lines. In contrast to primary cultures, cell lines reproduce indefinitely and contain an abnormal number of chromosomes. They grow much faster than cell strains.

"All cell cultures require the same growth factors and nutrients at approximately the same concentrations regardless of the animal from which the tissue is taken. This astonishing uniformity of metabolism is quite different from the requirements of bacteria and other microorganisms, which exhibit varied nutritional needs. Twenty-nine factors appear to be enough for supporting the growth of most cell cultures: 12 amino acids, eight vitamins, glutamine, dextrose or glucose, six inorganic salts, and serum protein. Compared with what an animal needs in its diet, a cell culture requires a greater variety of amino acids but fewer vitamins in its diet. The fact that a single medium can be used to grow a wide variety of cell strains and lines enables the experimenter to maintain many different kinds of cell for study.

"Before setting up a tissue culture the beginner should learn the elements of standard bacteriological procedures. [A good starting place would be your local university. Ed.] An autoclave is almost indispensable for sterilizing glassware, certain mediums and reagents. A large pressure cooker can serve as the autoclave. Materials placed in the autoclave will be thoroughly sterilized in 15 to 20 minutes at 121 degrees Centigrade, the temperature of steam at a pressure of 15 pounds per square inch. Start timing the sterilization after the pressure of the autoclave reaches this value. When sterilizing apparatus in the autoclave, always loosen the screw caps of bottles so that the pressure will equalize inside the bottle during the procedure. Cool the autoclave slowly, particularly after sterilizing fluids, to prevent the contents from boiling when returned to atmospheric pressure.

"Particles of dead bacteria, molds, and yeasts suspended in sterilized fluids can be removed by filtering the material through either filter pads or unglazed porcelain. An inexpensive and convenient apparatus for filtering consists of a syringe of the Luer type fitted with a Swinny adapter that holds the filter. A syringe of this type can be procured from the Fisher Scientific Company, *www.fischersci.com*, 4500 Turnberry Drive, Hanover Park, IL 60103, 800-766-7000.

"All reagents must be of the highest available purity. Triple-distilled water should be used in all procedures. The experimenter should anticipate spending a lot of time washing apparatus. All containers must be rinsed at least four times with tap water followed by a final rinse with triple-distilled water.

"For specimen materials beginners are urged to buy a starter cell culture. This material can be obtained from Difco Laboratories Inc., 920

Henry Street, Detroit, MI 48232, 800-638-8663. Information on available cultures, prices, and shipping will be sent on request. Alternatively, tissues and organs for culturing can be taken from an animal. This must be done under sterile operating conditions. [Note: Be sure to check the preparation instructions included in your kit. The following setup procedures may not be necessary if the materials are incubator-ready. Ed.]

"The fragments can be cultivated in either of two ways. One is to keep them in a form called 'plasma clot' on microscope slides or in flasks or tubes. The other is to disperse them as separate cells for cultivation in a liquid medium in tubes or bottles, where they grow as a single layer of closely spaced cells that adhere to the glass walls of the container. The choice between these two procedures depends on the use for which the cultures are intended. The plasma clot is excellent for microscopic examination but poor for maintaining cultures because nutrient can be made available to the growing tissue only a drop or two at a time. The nutrient must be replaced frequently. The monolayer type is widely used for perpetuating cultures, for investigating the interaction of cells and viruses and for studying cell lines.

"A plasma-clot culture is prepared by placing one drop of *TC-Chicken Plasma* in the center of a square of thin glass of the type used for covering microscope slides. To this drop is added one drop of *TC-Embryo Extract EE$_{20}$1*. (Reagents and mediums preceded by TC are products of Difco Laboratories. A numerical subscript indicates the percentage of extract or medium in the solution.) Mix the drops with a spatula and spread the fluid over an area about the size of a dime. Add two pieces of minced tissue to the center of the fluid. The specimens are now called explants. Each one should be about a millimeter square. Measure the size carefully. Explants larger than recommended cannot absorb adequate nourishment. Moreover, the initial size must be known so that the rate of subsequent growth can be determined.

"Cover the preparation and set it aside for about an hour, until it clots. An inverted microscope slide that contains a deep depression makes a convenient cover. Seal the cover glass to the slide with a ring of petroleum jelly. Incubate the preparation at approximately 37 degrees C. (98.6 degrees F.) and observe the culture under a microscope every 24 hours. The growing tissue must be transferred to fresh nutrient every two or three days, a procedure known as 'patching.' Simply cut the old slide culture back to a one-millimeter square, transfer it to a freshly prepared cover glass and continue incubation.

"I use a homemade incubator: a pair of nested cardboard boxes insulated with wool. Controlled heat is provided by a 75-watt lamp bulb regu-

lated by a thermostat of the type used in aquarium tanks. Such thermostats are also available with a built-in heating unit at slightly higher cost.
The temperature of the incubator should be maintained between 34 (93.2
degrees F.) and 37 degrees C.

"Cultures of the monolayer type involve several additional operations.
The cells grow as individuals. For this reason the term 'cell culture' seems
more appropriate than 'tissue culture.' To disperse the cells the fragments
of tissue are first placed in a flask containing a few glass beads, which
serve as agitators when the flask is swirled, and a sterile saline solution
that contains 0.25 percent trypsin. The trypsin dissolves the cement
between the cells to produce a suspension. Diluted trypsin does not affect
living cells and can be removed easily from the suspensions.

"To prepare the solution dissolve 25 milligrams of 1:250 trypsin (the
number is part of the name and designates the activity of the preparation)
in 10 milliliters of calcium-and-magnesium-free phosphate-buffered saline
(CMF-PBS). Filter-sterilize this preparation through a sterile Swinny filter
into half-ounce prescription bottles. [To obtain these materials, try Difco
Laboratories, the Fisher Scientific Company, or WARD'S Natural Science
Establishment (P.O. Box 92912, Rochester, NY 14692-9012; 800-962-2660
or *www.wordsci.com.*) Ed.] Store five milliliters in each bottle.

"To prepare the CMF-PBS solution dissolve 800 milligrams of sodium
chloride, 30 milligrams of potassium chloride, eight milligrams of sodium
orthophosphate mono-H, two milligrams of orthophosphate di-H and 200
milligrams of dextrose in 100 milliliters of triple-distilled water. Do not
autoclave this preparation but keep it frozen until you are ready to use it.
For dispersing tissue cells thaw the solution, add five fragments of tissue
of the same size as that used in the tissue-clot experiment, let the mixture
stand for six hours at 4 degrees C. and then shake it vigorously to make a
uniform suspension of cells.

"Now centrifuge the suspension at 800 revolutions per minute for five
minutes. Pour off the solution gently. Add five milliliters of basic salt solution and again shake the container to resuspend the cells. Filter the suspension through sterile gauze into a clean container and centrifuge it
again at 800 r.p.m. for five minutes. The washed cells collected at the bottom of the centrifuge tube are ready, after resuspension in basic salt solution, for monolayer culturing. Much of this work can be avoided, of
course, by buying prepared starter cultures.

"Mediums for making monolayer cultures are also available from
Difco Laboratories. I use Eagle's Basal Medium (also known as Eagle's
HeLa Medium). Glutamine must be added in the proportion of 30 mil-

ligrams per 100 milliliters of medium. Glutamine is unstable even at refrigerator temperatures; therefore it must be added as the medium is prepared for use. Make up the glutamine solution by dissolving 30 milligrams of reagent L-glutamine [available from Difco, Fisher, or WARD'S. Ed.] in two milliliters of triple-distilled water. Pass the solution through a Swinny filter into the container of medium.

"The medium must also contain serum—5 to 10 percent for maintaining a culture and 15 to 20 percent for encouraging growth. Serum provides growth factors that have not yet been identified. It also appears to encourage the attachment of cells to the glass walls. I use any of three serums: TC-Horse Serum, TC-Fetal Calf Serum or TC-Human Serum.

"To the medium thus completed antibiotics can be added for the control of bacterial infection. I use a combination of antibiotics that is effective against both gram-positive and gram-negative organisms. The combination consists of 100 units of sterile potassium penicillin G and 100 micrograms of dihydrostreptomycin sulfate per milliliter of medium. The antibiotics must be procured in the form of dry powders, without preservatives that might poison the cultures. A 500,000-unit vial of potassium penicillin G and a one-gram vial of dihydrostreptomycin constitute an adequate stock. The drugs can be obtained with the help of a cooperative physician.

"To start a monolayer culture, plant a quarter-milliliter of the cell suspension and one milliliter of the prepared medium in a screw-cap tube about 16 millimeters in diameter and 150 millimeters long. Close the cap tightly and place the tube in the incubator at an angle of about 15 degrees, so that the contents wet most of the wall at the bottom. Within 24 to 36 hours a dense monolayer will form where the fluid wets the glass. I substituted ordinary half-ounce prescription bottles for the screw-cap tubes. Bottles of this type have one flat side. (A druggist let me have six dozen for four cents a bottle.) The bottle is laid on its flat side. The flat inner surface appears to encourage the growth of exceptionally massive cultures.

"The growing culture exhausts the nutrient in about four days. To replenish the spent medium pour off the fluid, refill the container with four milliliters of fresh medium, and shake the solution gently until the monolayer disintegrates. Then transfer two milliliters of the fresh cell suspension to another bottle. This procedure is known as making a 'split.' You now have two cultures.

"The medium as supplied includes phenol red, an ingredient that serves to indicate the acidity or alkalinity of the fluid. The color of a fresh culture ranges from cerise to pink. As the cells metabolize, the solution gradually becomes acid, so that the color changes from pink to yellow. Alkalinity must then be restored by admitting air to the culture. This is done by loosening the cap of the bottle about a quarter-turn and retightening it.

"To stain monolayers for microscopic examination remove from a culture as much of the specimen as can be picked up with a sterilized wire loop approximately three millimeters in diameter. Transfer the cells to a clean microscope slide by pressing the loop on the glass. Wash the material gently with three changes of basic salt solution and then fix, or preserve, the cells by a drop of 10 percent formalin in 0.8 percent saline solution. Wash the material again to remove the formalin and let the specimen dry at room temperature. During the drying the cells become firmly attached to the glass; they can be stained by any of several preparations without becoming dislodged. Wright's stain is particularly easy to use. Flood the dried cells with two drops of a solution composed of 0.1 milligram of dry certified Wright's stain in 60 milliliters of acetone-free reagent methanol. [These materials can be obtained from Difco Laboratories, the Fisher Scientific Company, or WARD'S. Ed.] Let the stain act for two minutes and then add four drops of water. Let the slide stand for five minutes. Rinse off the stain and dry the slide at room temperature.

"An alternative procedure is to stain the cells by basic fuchsin—0.5 percent of the stain in a 20 percent solution of methanol in water. Let the preparation act on the cells for five minutes. Then rinse the cells in water or in a 50 percent solution of methanol in water. The diluted alcohol tends to produce slides of greater contrast between the cytoplasm and the nuclei than the water rinse does. On the other hand, the alcohol tends to bleach the dye. The slide must be watched carefully and the action stopped at maximum contrast by rinsing the preparation in water to prevent total destaining.

"As seen under the microscope monolayer cultures can be separated into two broad classes according to their appearance. Those that grow into long, spindly, loosely connected cells are called fibroblastic, because of their similarity to muscle fibroblasts. An example is the widely studied murine *L* cell line. Cells of the second class grow as closely joined polygonal shapes, called epithelial. The HeLa cell line is an example of this type, as are the great majority of other existing cultures.

"Because active tissues require frequent changes of medium, biologists sought a method of maintaining cells for long intervals with minimum attention. This was found in the technique of 'agar slant' culture, in which a supply of nutrient is in effect stored in an agar substrate. To establish a culture of this kind prepare a solution of Eagle's Basal Medium of twice the strength used for monolayer cultures. Next make up a 3 percent solution of

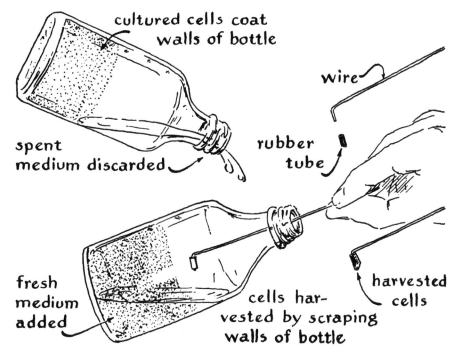

Figure 15.1   Harvesting monolayer tissue culture.

Noble agar ('Noble' indicates a purified grade of agar) or of ordinary agar washed five times in cold, triple-distilled water and then dried. Autoclave the solution and cool it. Mix equal volumes of the concentrated medium and agar solution, dispense the mixture in four-milliliter amounts in sterile test tubes and cool the tubes in an almost horizontal position so that the agar solidifies as a 'slant' that extends from the bottom of each tube almost to the top. When the agar has cooled, pipette 0.3 milliliter of Eagle's Basal Medium with 10 percent TC-Horse Serum into the tube. Store the tube upright in the refrigerator. Such slants are satisfactory for six months or more of use.

"In order to establish a tissue culture in the agar slant tubes, harvest cells from a young monolayer (two or three days old) and then wash them once with sterile basic salt solution *[see Figure 15.1]*. Separate the cells from the fluid by centrifuging. Gently pour the solution from the centrifuge tube without disturbing the cells that have settled to the bottom. Add half a milliliter of sterile basic salt solution to the tube and agitate it to make a dense suspension of cells. Using a wire loop that has been sterilized by flame, transfer a loopful of the cell suspension to a prepared agar slant, touching the agar with the loop at several places. (Do not smear the inoculum by dragging the loop across the surface of the agar.) Cap the agar tube tightly and incubate it in an upright position at 37 degrees C. Colonies of cells will appear in about four days and will grow to a diameter of five to 10 millimeters within 10 days. The nutrient solution should be changed at least every three weeks. The culture will live for six weeks or longer. For this reason agar slants are well suited for keeping stock cultures of cell lines and cell strains."

16 EXPLORING SLIME MOLDS

by C. L. Stong, January 1966

You can usually stump players in guessing games such as Twenty Questions by introducing into play one of the 450 species of amoeboid slime molds. These are organisms that do not fit neatly into the conventional categories of animal, vegetable, or mineral. Strictly speaking they are none of these. Some player may argue that the zoologist classifies the organisms as animals—protozoa in the group of animals called Mycetozoa—on the basis that they cannot manufacture food and that during much of their life cycle they move freely in response to light, heat, and chemical stimuli. To this argument the botanist can respond just as stoutly that slime molds are plants, which he calls Myxomycophyta, because at maturity the molds develop cell walls of cellulose, become securely attached to some feature of the environment, produce fruiting bodies, and release spores.

Some biologists take the position that the concept of plant or animal simply cannot be applied to these organisms. At some point in the course of evolution the slime molds ran up a blind alley, where they are now trapped—relatively harmless and apparently good for nothing except as specimens for some of the most engaging experiments in biology. A. C. Lonert, who is director of research at the General Biological Supply House in Chicago, explains how to culture and experiment with two of the most interesting species. "I shall discuss two kinds of slime mold belonging to different orders," he writes, "because of the astonishing diversity of their life cycles and the ease with which they can be cultured. Both organisms offer a rich field for experimentation.

"The first, *Physarum polycephalum*, is a member of an extremely interesting order, the widely distributed Myxogastrales (also called Myxomycetes) of the subclass Mycetozoa. Viewed from the perspective of an imaginative microscopist, the sporangia, or fruiting bodies, of the many species appear as enchanted forests in which the fruits become pulsating, devouring monsters.

"When found in the early stage of growth, the myxomycete is most like an animal and generally consists of an unattractive patch of naked protoplasm, either colorless or pigmented, that can attain a diameter of several feet. This hungry protoplasmic mass, called a plasmodium, creeps about in a moist and well-shaded environment, seeking food in rotting vegetation and decomposing wood. In the process of feeding, the organism aids the economy of nature by processing waste into compounds that can be utilized by other organisms.

"At a propitious time, governed by the availability of food and such factors as temperature and moisture, the oozy mass leaves its dank haunts and, as it reaches toward the light, transforms itself into upright plantlike structures. These are the sporangia. The shape, color, and size of these bodies vary considerably among the many genera of Myxogastrales, but the bodies usually have two features in common: a delicate network of noncellular threads and strands, called the capillitium, in which the spores are enmeshed, and a fragile, sometimes beautifully formed and tinted spore case, called the peridium.

"After the drying and rupture of the spore case, expansion of the threads of the capillitium exposes the spores to the action of wind and rain. At the same time the expansion controls the rate at which the spores are released. The life cycle is repeated when spores fall in a suitable environment. Free-swimming swarm cells, or myxamoebae, are then released. They feed for a time and multiply by splitting in two. Eventually two of the cells fuse to form a zygote, which is equivalent to a fertilized egg. This fusion constitutes the beginning of a new plasmodium, which can grow by nuclear division or by fusion with other zygotes. Unfortunately for those who enjoy the exotic and beautiful in nature, these fantastic groves are usually at ground level. Because they are underfoot they usually remain unobserved and unappreciated except by those who know where to look.

"Most of the slime molds, when taken from their natural environment, are difficult to culture. This difficulty need not stop anyone from enjoying the real-life adventure of observing the strange organism at home or in the laboratory. A culture can be started easily from a fragment of *Physarum polycephalum* plasmodium in the sclerotial stage, a dried condition. This stage is sometimes found in nature and can be induced in the laboratory by the method to be described.

"Sclerotia that will remain viable for a long time and have the ability to transform themselves into active plasmodium within several hours are available commercially. One can also buy the necessary supplies for experimenting with the organisms. All the items can be obtained from distributors such as WARD'S Natural Science Establishment (P.O. Box 92912, Rochester, NY 14692-9012; 800-962-2660 or *www.wardsci.com*).

"To grow a plasmodium of *Physarum polycephalum* from the sclerotial stage, first improvise a culture chamber by placing a dish—for example an inverted saucer or a petri dish—within a larger, covered container, such as a casserole or a pair of facing Pyrex pie plates, so that the smaller dish serves as a platform *[see Figure 16.1]*. For this experiment omit the rubber separators shown in the illustration. Lay a sheet of filter paper or paper toweling across the platform. The paper should droop to the bottom of the dish. Wet the paper thoroughly with distilled water, pouring off the excess.

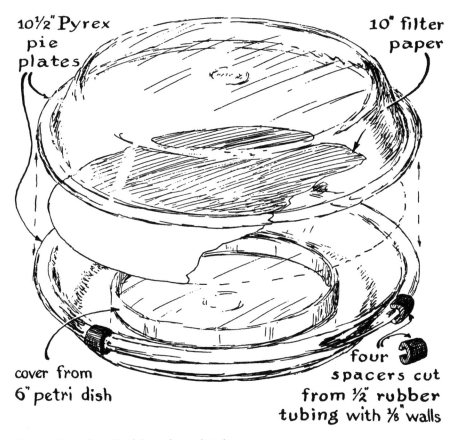

Figure 16.1 Details of the culture chamber.

From time to time add enough water to prevent drying. Incidentally, distilled water should be used in all experiments to be described.

"Place a small fragment of sclerotium on the platform and wet it with a drop of water. Within a few hours the organism will awaken from its deathlike torpor. Keep the unused portion of sclerotium in the refrigerator. When the plasmodium has emerged and has begun to seek food, put a flake of uncooked oats in contact with the rapidly spreading growth. Use old-fashioned rolled oats, not the 'quick' variety.

"Keep the covered dish at room temperature in an area that does not get direct sunlight. As the plasmodium increases in size, place more oat flakes along the growth front. A rapidly moving plasmodium will consume a larger number of oat flakes than a sluggish one. In either case the feeding organism will show a decided preference for fresh oat flakes and will abandon partially digested ones. Flakes that have first been moistened with a drop of water are accepted more readily than dry oats. To maintain a clean culture, transfer the organisms to a fresh sheet of filter paper or paper toweling weekly, avoid overfeeding, and remove abandoned food that shows signs of becoming moldy or slimy. Occupied oat flakes (those covered with yellow plasmodium) or sections of paper bearing plasmodium can be transferred repeatedly to clean sheets of paper to perpetuate the plasmodial stage.

"The organism at this stage has been compared to a giant amoeba: a thin, irregularly shaped mass of creeping jelly that is threaded by a network of veins. To observe the complex circulation in these vessels, transfer several occupied oat flakes to a petri dish containing a thin layer of nonsterile, nonnutrient 1.5 percent agar. The agar culture is maintained in the same way as the paper culture, but there is no need to add water. The protoplasmic streaming and its periodic reversal can be observed with any microscope capable of magnifying to 50 diameters or more. Observation should be made with light transmitted through the specimen from the substage. Cut one of the vessels when the circulation is active and observe the effect. Make a transfer of scraped plasmodium and watch it reconstitute itself. Place a drop of vinegar near the plasmodium and note the reaction.

"Usually cultures can be kept in the plasmodial stage for some weeks by maintaining ample food and water and periodically transferring specimens as described. Occasionally, however, specimens will for some unknown reason enter the sporangial, or fruiting, stage spontaneously. This stage can be induced at any time by removing most of the food and allowing the plasmodium to roam while simultaneously keeping the organism moist. The transformation will occur suddenly, usually at night, within a week or so. An observer who is fortunate enough to witness the actual transformation will see the entire plasmodium, now more orange than yellow, appear to separate into uniformly rounded masses that ascend from the surface on stalks and then develop into weird, multilobed bodies.

"Under the higher powers of the microscope, spores obtained from crushed sporangia can be seen to germinate with the emergence of an amoebalike protoplast, which afterward gives rise to an amoeboid and then to flagellum-bearing swarm cells—cells equipped with whiplike tails for swimming. The series of transformations usually requires about three days. Eventually pairs of swarm cells fuse to become zygotes. From these a new plasmodium originates.

"The germination and subsequent transformations up to, but usually not including, the plasmodial stage can be observed by setting up a hanging-drop preparation. This consists of a drop of water, containing specimens, that clings to the underside of a thin sheet of glass. The glass is usually a cover slip of the type used to protect specimens on a microscope slide. Evaporation is prevented by enclosing the drop in a small, airtight vessel. The vessel may consist of a microscope slide into which a depression has been ground. The other items needed for doing the experiment are a sterilized medicine dropper, a quarter-inch loop of fine wire supported at the end of a pencil-sized piece of wood, a pair of tweezers, Vaseline, a 70 percent solution of isopropyl alcohol, and a gas burner or an alcohol lamp.

"Sterilize a microscope slide and cover slip in the flame. A few sporangia are then placed in isopropyl alcohol for one minute and dried on the sterilized slide. Next the dried specimens are placed between another pair of sterilized slides with a drop or two of water and crushed. The sterilized cover slip is held horizontally, by an edge, with the tweezers. A drop of water is applied to the underside of the cover slip at the center.

"This suspended drop is inoculated with spores by first stirring the crushed specimen with the wire loop and touching the loop to the bottom of the drop. The rim of the depression in the microscope slide is now lightly coated with Vaseline. Finally, the cover slip is placed on the coated slide with the suspended drop centered over the depression. The cover slip is pressed gently into place to seal with Vaseline the space between the cover slip and the slide. The combination is placed on the stage of the microscope, which is focused on the contents of the suspended drop. The recommended sterile technique is not absolutely necessary, but it increases the probability that more of the early stages will develop. It is difficult, however, to generate a fresh plasmodium by this technique.

"To preserve *Physarum polycephalum* for possible future use, return the organism to the dormant, sclerotial stage, from which it can be conveniently aroused again when it is needed. A method that I developed is quite simple and produces a good yield of very durable sclerotia. Set up a culture chamber with 10½-inch Pyrex pie plates, 10-inch disks of paper and either the top or the bottom of a six-inch petri dish to serve as an elevated platform. Lay the paper across the platform and center it on the lower pie plate. Moisten the paper with distilled water and push down the portion of the paper that extends beyond the platform. The depressed paper forms a circular moat. Pour off excess water. Transfer oat flakes occupied by plasmodium from a desired culture to the center of the paper. Cover the pie plate with a similar but inverted dish.

"When the plasmodium begins to move about, place fresh oat flakes along the growth front—wetting each flake to facilitate rapid occupation. By the following day plasmodial fronts, moving toward the edge of the culture chamber, will have developed. Put fresh oat flakes along these more distant fronts and, when they are occupied, lift them from the paper with a knife, replacing them promptly with fresh flakes, and reposition the occupied flakes toward the center of the platform. Place the flakes next to each other. Continue this procedure until a circular area of occupied flakes about three inches in diameter has been concentrated centrally on the platform. As much specimen material as desired can usually be gathered after 48 hours of consecutive feeding. Occupied flakes can be taken from more than one culture; if several cultures are maintained simultaneously, a good supply can be accumulated in a much shorter period of time.

"To begin the induction of sclerotization, take the plasmodium-bearing paper out of the culture chamber. Remove surplus moisture by superposing the paper gently, with occupied oat flakes uppermost, on a piece of clean, dry paper of the same size. Allow the papers to remain in contact. Wash and dry the platform and bottom culture plate, then promptly reassemble the chamber and restore the partly drained culture paper to its former position, re-forming the moat.

"Circle the paper moat with dry oat flakes. The flakes form an entangling barrier that prevents wastage of active plasmodium by premature drying at the edge of the plate. Next insert one or two thicknesses of rubber tubing, which have been slit lengthwise, between the upper and the lower culture plates at four equidistant points *[see Figure 16.1 on page 121]*. If the relative humidity is high (from 50 to 60 percent at temperatures ranging from 80 to 90 degrees Fahrenheit), use two thicknesses of rubber tubing and increase the spacing of occupied flakes; if the relative humidity is low (from 15 to 25 percent at 75 to 80 degrees F.), use one thickness of rubber tubing. These are, of course, rough approximations. The object is to dry the culture paper in not less than 24 hours and not more than 48 hours. (One should avoid, however, the use of an artificially created air current to promote drying.) The crustlike sclerotia produced by this method will retain viability for years if they are stored in closed jars under refrigeration at 42 degrees F.

"The second slime mold, *Dictyostelium discoideum,* a member of the order Acrasiales, differs in a number of remarkable ways from *Physarum polycephalum* and other members of that order. It is a bacterial feeder and also exhibits both animal and plant characteristics. The events of its life cycle can be ascertained rather easily, but several of its other properties have not yet yielded to the assaults of the investigator. Apparently similar cells play at least three distinctive roles on a simple level—first, as free-living myxamoebae; second, as massed structures of cells exhibiting interdependence and coordination; and third, as cells differentiating into several kinds of structure.

"About 18 to 24 hours after inoculation of a bacterial culture with spores of *Dictyostelium discoideum* there will appear within the bacterial growth numerous minute, refractive lumps that have bright centers when they are viewed slightly above focus at a magnification of 100 diameters. These lumps are actually the minute myxamoebae busily eating their way through what, on the scale of comparative sizes, must be described as jungles of bacteria. Simultaneously the organisms multiply at a prodigious rate through binary fission, or splitting. Even a hand magnifier will show the digestion of circular patches of bacterial growth at this stage. Incidentally, if the experimenter is working with an inexpensive microscope objective of 0.25 or 0.65 numerical aperture (corrected for use with a cover slip) and a 160-millimeter tube length, a fairly good compensation for the aberration caused by the absence of the cover slip can be made by increasing the tube length by 30 millimeters.

"At the end of 24 hours the culture will begin to teem with hordes of the voracious organisms, which are now clearly visible as they sweep the agar surface clean of bacteria.

"By the end of 48 hours, with the food supply nearing depletion in some portions of the plate, a remarkable change will begin to take place. The organisms stop feeding and spontaneously converge on centers of aggregation. According to John Tyler Bonner of Princeton University, the release of acrasin, a chemotactic substance, is responsible for the strange phenomenon. As the myxamoebae continue to press themselves into these centers of aggregation several moundlike structures are built up, each with a characteristic tip. The myxamoebae pack themselves together but persist in maintaining their individuality. There is no fusing of cells into a mass that contains a large number of nuclei as there is in the case of *Physarum polycephalum;* the pseudoplasmodium remains associational in character. The individual aggregates grow rapidly through the continued accretion of myxamoebae, elevating at first and then, after a period of elongation, tipping horizontally to become migrating, sluglike creatures.

"The slug soon assumes the characteristics of a creature equipped with special organs. For example, it has a front-to-back orientation, with the tip, observed to form early in the mound stage, exhibiting sensitivity to light and heat. In the presence of such a stimulus the top appears to head the parade with the rest of the body following, leaving a trail of slime behind it. The motion of the slug is apparently produced by the concerted action of the individual myxamoebae, but it is not yet known what coordinates their remarkable performance. It is possible, by exerting gentle pressure, to dissociate amoebae from the mass. Subsequently they recombine to reconstitute the slug! Even a stained section of the slug fails to demonstrate the mechanism of communication. It is at this stage of development that the slime mold lends itself particularly well to experimentation.

"Bonner has shown that when the tip of one slug is cut and joined to the tail end of another slug, or replaces the tail of another slug, it becomes evident that differentiation has occurred, that certain groups of the amoebae have acquired unique traits. Myxamoebae of the tip that have been transplanted, for example, will begin to migrate to their own social level and join the tip cells of the host. On the other hand, a graft of the hind part of one slug to replace the hind part of another slug produces no appreciable migration. To tag the myxamoebae for these and other experiments Bonner stained the specimens with a harmless dye. The stains must be used in highly diluted form; they include Janus green B, methylene blue, neutral red, and brilliant vital red. Pink slugs can also be developed from colorless slugs for experimentation by starting a culture with the red *Serratia marcescens* bacterium as the food organism. The myxamoebae are unable to eliminate the bacterial pigment.

"The final stage should develop in about 60 hours. By that time a number of slugs will have begun to elevate and change into fruiting bodies. In the process the myxamoebae located toward the front end and along the central axis of the slug undergo a drastic change, literally giving up their lives to become a firm supporting core of cellulose-lined dead cells constituting the sporophore, or stalk. The rest of the myxamoebae begin to be raised upward. Those amoebae that do not become stalk cells are eventually lifted in a body to the top of the stalk, where they are transformed into capsule-shaped spores encased in a globular mass of slime. This body constitutes the sorus. The complete fruiting sequence is shown in the accompanying illustration *[see Figure 16.2]*. If spores are sown on a clean agar surface without food organisms, freshly hatched myxamoebae, the beginning of a new cycle, may be observed in about 18 hours.

"A suitable culture medium can be made by first cooking 35 grams of hay for 10 minutes in one liter of tap water. Filter this solution and add enough water to restore the original volume. Add the filtered infusion to a flask containing 15 grams of agar. Boil to dissolve the agar. Sterilize in an autoclave [an ordinary pressure cooker will work well here. Ed.] at 15 pounds of pressure for 20 minutes.

"Another satisfactory medium consists of 0.1 percent lactose and 0.1 percent peptone in 2 percent agar sterilized as above. Melt the agar first and then add the other ingredients. To start the culture pour the medium into sterile petri dishes and test tubes. Avoid aerial contaminants. Test-tube media are usually sterilized after pouring. After preparing fresh stock cultures inoculate the selected medium with a nonmucoid strain of *Escherichia coli* or *Serratia marcescens* [available from WARD'S Natural Science Establishment. Ed.] and then inoculate the same plate with *Dictpostelium discoideum* spores. By restricting to limited areas the spore inoculation of the petri-dish culture (the entire plate, however, can be inoculated with bacteria) almost any stage of development can be found in other portions of the culture for several days after formation of the first organisms. The bacteria that serve as the food organism can be

Figure 16.2 Stages in the fruiting of *Dictyostelium discoideum*.

transferred in the form of a suspension or they can be streaked over the agar surface. Incubate the culture at room temperature.

"To prepare stock cultures transfer the food organism to two or more agar plates and add *Dictyostelium discoideum* spores to one of them. When the stock culture has developed, stopper it with the separate culture of bacteria and store under refrigeration. Transfer the stock cultures every three months."

17 FABULOUS PHOTOTAXIS

by C. L. Stong, October 1964

W hy does a moth fly into the flame of a candle, a cricket scurry for dark cover when exposed to light by the overturning of a rock? Such behavior, known as phototaxis, is observed in many organisms, plants as well as animals. Although a century of investigation has failed to explain the complex mechanism of phototaxis, the work has produced a number of interesting experimental techniques useful in the more general study of how living things react to their environment. Techniques of this kind have been particularly useful in the study of the single-celled organisms that exhibit marked response to changes in illumination. Recently, Werner C. Baum, associate professor of biology at the State University of New York at Albany, developed a series of experiments for investigating phototaxis; amateurs can easily adapt them for demonstrating how variations in age, nutrition, temperature, and light intensity may alter the response of single-celled organisms to light of various colors.

"Light," Baum writes, "is an immediate environmental factor in the life of most organisms. In the case of plants one thinks chiefly of photosynthesis, but the less overt processes of phototropism (plants turning toward light) and photoperiodism (the response of plants and animals to variations in the relative duration of day and night) are also controlled by light. The common denominator of these varied effects of light is the absorption of light energy in one or more photochemical, temperature-independent reactions and the conversion thereby of light energy into the energy of chemical bonds.

"In the case of phototaxis, the entire organism moves in response to light. Positive phototaxis is exhibited by the flight of insects toward a light source at night and by such behavior patterns as the vertical layering of zooplankton in surface waters during the day. Negative phototaxis is exhibited by darkling beetles and their larvae burrowing under the oatmeal on which they are being cultured and by the myriad of arthropods living under a log or rock.

"My interest in phototaxis was stimulated by the microscopic, single-celled organism called *Euglena*, which propels itself through water with a whiplike flagellum. *Euglena* has chloroplasts and can carry on photosynthesis, so that it qualifies in this respect as a plant. It also is capable of growing on appropriate media in the dark, so that it qualifies in this respect as an animal. Its high degree of motility, anterior gullet, contractile vacuoles, and its external, probably noncellulosic pellicle, or exoskeleton, are 'rule of thumb' animal characteristics *[see Figure 17.1]*.

"It is likely that generations of biology teachers have plagued or stimulated their students by asking them if *Euglena* is a plant or an animal in the hope that they will perceive from this loaded question the arbitrary nature of biological classification at certain levels and the continuity of morphological and physiological features in various organisms. The characteristics of *Euglena* combine to make it not only an interesting and instructive organism but also an excellent experimental subject.

"One of the more conspicuous features of most species of *Euglena* as observed under the microscope is an orange-red body called the stigma, or eyespot. In the electron microscope it appears as a loose aggregation of about 50 granules. The color of the eyespot is due to one or more carotene derivatives, the exact identities of which are still in doubt. As a group these pigments are involved in a number of photochemical processes in both plants and animals. Evidence suggests that the eyespot is closely related to

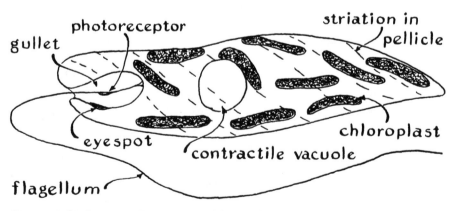

Figure 17.1   Structure of the organism *Euglena*.

phototactic responses in *Euglena* as well as in other organisms, although phototactic responses occur in organisms without eyespots.

"In addition to the eyespot a dense nodule near the base of the flagellum on the side facing the eyespot is variously designated as the photoreceptor and the paraflagellar body. Some investigators consider it of major importance in the mechanism of the phototactic response. According to one concept, positive phototaxis is related to a periodic darkening of the photoreceptor by the eyespot as the *Euglena* moves forward in its spiral path. This active orientation and swimming of *Euglena* in relation to the light source is designated as topophototaxis. It is distinguished from the phobophototaxis of certain bacteria: a spontaneous return to a region of greater light intensity whenever the light intensity elsewhere diminishes. Thus the organism is 'trapped' in the light spot. The phototaxis of some organisms may involve both mechanisms.

"Although experiments on phototaxis using *Euglena* predate the work of the German biologist T. W. Englemann, it was he who in the 1880s carried out classical experiments on both photosynthesis and phototaxis. His experiments have served subsequent investigators as models of ingenuity and experimental design. Englemann worked with a purple sulfur bacterium, with *Paramecium ursaria,* and with *Euglena* as well as with other organisms. He correlated the phototaxis of *Euglena* with the eyespot. In the purple sulfur bacteria he demonstrated both phototaxis and a correlation of phototaxis with photosynthesis in the near-infrared portion of the spectrum. He made an inference, subsequently confirmed, that in photosynthetic bacteria light of a color that most strongly attracts the organism also promotes photosynthesis most effectively. The energy involved in both processes is absorbed by the same pigment system.

"Investigators since Englemann have concentrated on pinpointing the absorption and action spectra and on elucidating the mechanism of the phototactic response, particularly the functions of the eyespot and the photoreceptor in organisms that have those parts. The work also includes studies of phototaxis in organisms that lack eyespots, such as the purple sulfur bacteria, desmids and blue-green algae, and of other organisms that possess eyespots, such as *Chlamydomonas* and *Volvox*. Comparative studies of phototaxis have been made on strains of *Euglena* with chloroplasts, eyespots, and photoreceptors; on strains without chloroplasts but with eyespots and receptors, and on strains with chloroplasts and photoreceptors but without eyespots. Chlorophyll-free *Euglena* can be prepared by the bleaching action of certain chemicals such as streptomycin. Strains without eyespots have been obtained by subjecting specimens to ultraviolet radiation.

"Phototaxis in *Euglena* is frequently illustrated by exposing a culture in a transparent container to a beam of light. The active organisms promptly move to the side of the container nearest the light source, where, even though individually microscopic in size, they congregate in such numbers as to be readily apparent to the naked eye. It occurred to me that the effect might be emphasized by completely darkening the culture except for a small spot on the side wall of the container. I covered a culture contained in a 100-milliliter glass jar with a mask made of black paper. A single ¼-inch opening was punched in the paper mask. When the culture was exposed to light, a dense aggregation of *Euglena* collected near the opening in less than an hour.

"To demonstrate the influence of color I next made a black paper sleeve with windows covered by transparent plastic of various colors. A series of six ¼-inch holes ⅜ inch apart was punched in a piece of black construction paper with an ordinary paper punch. A small strip of clear or colored cellophane was placed over each hole and secured with cellophane tape *[see Figure 17.2]*. Dennison-packaged DuPont cellophanes, available in different colors, were used as crude filters. [You should be able to obtain these cellophanes at a stationery store. Ed.]

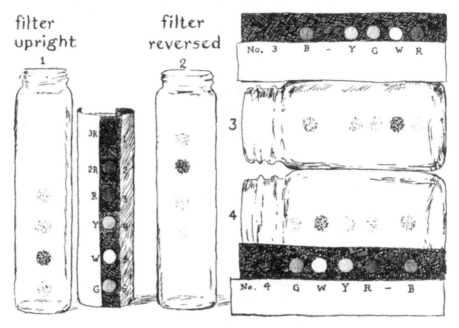

Figure 17.2   Effect of various color filters.

"The assembled device was carefully fitted to a glass vial so that the interior was completely darkened with the exception of the colored windows. The vials were about 3½ inches high and about an inch in outside diameter; both round and rectangular types were used. The fitted paper sleeve was assembled with masking tape so that it could be slid from the vial without agitating the culture.

"A *Euglena* culture was placed in a vial, the vial was stoppered, and the window side of the fitted sleeve was exposed to a beam of white light. Almost any light source was found to be satisfactory. A fluorescent lamp is weak in red wavelengths. When one uses a tungsten lamp, one should take care to assure uniform distribution over the exposed side of the culture vessel, and the lamp should not be so close that it heats the vial. Light intensity can be controlled in part, and the effect of intensity differences can be studied, by placing cultures at various distances from the light source.

"The approximate intensity of the light passing through the color filters at the distances involved (and presumably reaching the organisms in the vial) was measured by a light meter. Meters of the type used for determining photographic exposures can be used. The clear filter transmitted the most intense beam, followed by blue, then yellow, green, and red in order of decreasing intensity. The approximate corresponding range of wavelengths was determined by measuring sample strips of the colored plastic in a colorimeter that indicated the percentage of light transmitted. The light turned out to be far from monochromatic. For example, the blue plastic transmitted blue light as follows: at a colorimeter setting of 400 to 450 millimicrons (blue), 90 percent; at 550 millimicrons (green), 60 percent; at 650 millimicrons (red), 23 percent. For a given range of wavelengths the intensity was varied by using two or more thicknesses of the filtering material. Complete data on the light-transmission characteristics of the plastic material are usually available from the manufacturer.

"Active *Euglena* cultures moved to the colored windows in less than 15 minutes. White light produced the quickest response, followed by blue, green, yellow and red. The number of organisms that assembled at each window was also greatest in the case of white light and declined in the same sequence: blue after white, then green, yellow, and red. The sequence was the same whether the culture was lighted for only a few hours or overnight. The maximum phototactic response in blue light coincides with the maximum absorption of blue light by the eyespot. If aggregation of the organisms is allowed to continue for periods of more than 30 minutes, the *Euglena* adhere to the walls of the glass vials even when the culture is mildly agitated. The cultures can then be carefully poured off and the vials inverted on a paper towel for drying. After drying, the *Euglena* aggregates can be fixed to the glass by placing the vial next to a low heat source for a few minutes.

"A more versatile color filter was next constructed in which different-colored plastic strips ⅝ inch wide were placed across rectangular slots, measuring about two inches by ¼ inch, cut in a black paper cylinder. The strips were arranged horizontally and overlapped about ⅟₁₆ inch. They were held in place by strips of gummed paper placed along the overlapped edges. Still a third variation of the filter assembly was prepared by securing narrow strips of colored plastic across a wide slot (about 1¼ inches) cut in a 3-by-5-inch card. This proved to be a very flexible device, easy to place over slots in opaque sleeves that fit a variety of containers in assorted shapes and sizes [see Figure 17.3].

"The results of experiments made with the color filters were so encouraging that I decided to subject cultures to an actual spectrum in the hope of observing the sharpest possible differentiation of phototactic response with respect to the wavelength of light. A simple apparatus for dispersing the light was improvised from materials that were at hand. The rays of a 60-watt incandescent lamp were refracted into a beam of parallel rays by means of the lens from a reading glass and were dispersed into the spectral colors by a glass prism that measured 1½ inches by two inches. The colors were then

Figure 17.3   Evolution of an experiment with color filters.

projected onto a culture of *Euglena*. A pronounced differential response was immediately apparent. Maximum aggregation occurred in the red area of the culture! This was contrary both to the results previously observed and to the literature. The puzzle was resolved by measuring intensity across the spectrum: the light meter indicated that the red portion was almost 20 times more intense than the blue. The difference in energy was therefore masking the effects of the difference in wavelength."

(If the amateur does not own a prism, an alternate source of intense spectral light can be improvised by equipping an ordinary 35-millimeter projector with a vertical slit in the position normally occupied by the slide and placing a replica diffraction grating in front of the projection lens. This scheme is suggested by Roger Hayward, who illustrates this department. Transmission gratings of adequate quality for this application can be obtained, in a sheet that measures 8 by 11 inches, from the Edmund Scientific Company in Tonawanda, New York. Details of the arrangement are depicted in *Figure 17.4*.)

"On one occasion, when a very dense, older culture of *Euglena* was exposed to light overnight, I was surprised to find no phototactic response. Although apparently all experimental conditions were in order, the organisms failed to aggregate and to adhere to the side of the vessel. This and

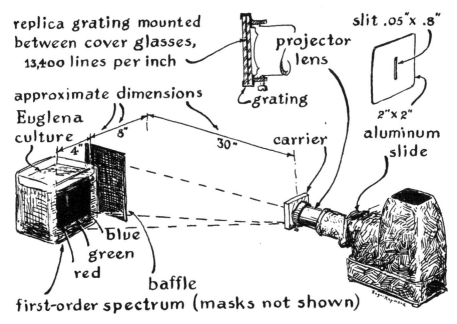

Figure 17.4 Apparatus for splitting white light into the spectrum.

related evidence suggested that not all cultures are phototactically equivalent. Therefore care should be taken in the design of experiments to equate only the results of cultures that are comparable in terms of age, nutrition and similar factors.

"Since there is some indication that the inorganic environment may influence phototaxis (and other processes as well) I decided to investigate whether or not this might be a factor in the lack of response of the inactive cultures. The inactive organisms were older, denser cultures in which the pea extract I used as a nutrient in the medium had not been replenished for several weeks. In lieu of waiting for a culture to become nutritionally depleted one could separate the organisms from the culture solution by aggregating active *Euglena* on the sides of a vessel and either pouring off the balance of the culture, centrifuging, or filtering. The aggregated organisms could then be resuspended in tap water or distilled water.

"To investigate the effect of mineral nutrition I dissolved a commercial N-P-K (nitrogen-phosphorus-potassium) plant-fertilizer tablet in water and added the equivalent of half a tablet to a 400-milliliter suspension of phototactically inactive *Euglena*. I kept another suspension without fertilizer as control. Essentially no phototaxis and aggregation were observed in the control, but the experimental culture to which the plant tablet had been added displayed marked activity. This suggests that mineral nutrients may be a decided factor in phototaxis.

"In another experiment made to disclose the effect of mineral salts on inactive organisms, nitrogen, phosphorus, and potassium were added to separate phototactically inactive cultures in amounts equivalent to their concentration in the plant-fertilizer tablet. To each 100 milliliters of cultures of inactive *Euglena* I added 15 milligrams of phosphoric acid, 16 milligrams of potassium chloride, and 190 milligrams of ammonium sulfate. Light intensity was maintained at 400 footcandles. The light source was placed 20 centimeters from the cultures. The light exposure was five hours. The salts most effective in promoting phototaxis and aggregation were, in descending order, phosphorus, nitrogen, and potassium. All produced more aggregation and phototaxis than appeared in the control. These results indicate not only the general effect of mineral nutrients on phototactic response but also the differential effects of the several ions, particularly the pronounced influence of phosphorus and nitrogen.

"Sets of inorganic 'sufficient-and-deficient' plant-growth salts are available from biological supply houses. The results of experiments made with them not only are readily apparent to the eye but also are automatically plotted by the differential adhesion of the organisms to the glass. Permanent records of the responses can be made by simply photographing the aggregated organisms. The same general procedure can be used for investigating the effect on phototaxis of light intensity, pH, drugs, vitamins, hormones, and age of cultures.

"Active *Euglena* organisms as well as many other cultures are stocked by the larger biological supply houses. Pure cultures of both green and colorless species of *Euglena* for research uses are available from WARD'S Natural Science Establishment; 800-962-2660 or *www.wardsci.com*. Algae and other organisms can also be collected in the field. The specimens are placed in an aquarium, concentrated phototactically by a light source placed at one end of the aquarium in an otherwise darkened room, and pipetted into a simple culture medium. Alternatively, the organisms can be placed in a volumetric flask that is darkened except at the neck; after they have aggregated they can be pipetted out of the neck.

"Two nutrient media are widely used for culturing *Euglena*. The split-pea medium is made up of the fluid obtained by boiling 40 split-pea halves in one liter of tap or pond water for a few minutes and then discarding the solid residue. The second medium, known as the soil-water type, is also highly recommended for *Euglena* as well as for a wide variety of other algae. It is prepared by adding successively to a test tube a pinch of calcium sulfate, a half-inch of good garden soil and a quarter of a dried split pea. About 75 milliliters of water (tap, pond, or distilled) is then added along the side wall of the test tube. After the test tube has been loosely plugged with cotton it is steamed (do not autoclave) for one hour on each of two successive days. After the resulting fluid has cooled and cleared by settling it can be inoculated with the desired organisms.

"I have used split-pea medium with consistently satisfactory results. One may observe a temporary rapid increase in bacteria after adding split-pea medium to a culture. The bacteria will diminish over a period of several days as the concentration of the Euglena increases until they are no longer apparent.

"The experiments need not be confined to *Euglena*. Although most species of *Paramecium* will be found to be indifferent to moderate light intensities, *P. bursaria*, which plays host to enough green algae to give it a green and plant-cell-like appearance, is positively phototactic. In these organisms phototaxis has been found to be a response related to the oxygen produced by the symbiotic algae in photosynthesis. It is sometimes referred to as secondary phototaxis arising from chemotaxis or aerotaxis. I once observed the response by placing a culture of *P. bursaria* in a miniature beaker that was completely darkened except for a single ¼-inch hole punched in the black paper on the bottom. The culture was placed on the stage of a stereoscopic binocular microscope equipped with a transilluminating substage. The hole was illuminated overnight from below, the spot of light being centered in the field of view. By morning the spot was covered with *P. bursaria*.

"In a similar experiment made with five green *Hydra* only three organisms were found in the light spot after an overnight exposure. Incidentally, the prior condition of organisms tends to influence their response to a given set of conditions. In my experience more uniform patterns are observed when the organisms are kept in darkness for a 24-hour period before their use in an experiment.

"Many refinements of the above techniques can be developed for investigating the phototactic responses of an entire range of smaller organisms. Given a choice of various wavelengths, where would *P. bursaria* and *Hydra viridissima* preferentially aggregate? Flatworms are considered negatively phototactic. If given no opportunity to remain in the dark, in what wavelength, if any, would they preferentially remain? Many smaller crustaceans exhibit phototaxis; brine shrimp, for example, are readily cultured in the laboratory. What effect, if any, would a competing population of other organisms have on their phototactic responses?

"The fundamental question of just why and how the energy of light triggers and then guides the swimming motion of these organisms still awaits explanation. Doubtless the full answer will come when data from experiments such as these are correlated with comparable information derived from the disciplines of cellular physiology, biochemistry, biophysics, and electron microscopy."

PART 4

ENTOMOLOGY

18 LOVELY LEPIDOPTERA

by Albert G. Ingalls, October 1954 and Shawn Carlson,
July 1997

At the age of 73, Colonel Otto H. Schroeter of Quaker Hill, Connecticut, is still chasing butterflies. Except for time out while he studied engineering as a young man in his native Germany, he has been at it more than six decades, including the years when he was employed as construction superintendent in the Panama Canal Zone by the Isthmian Canal Commission.

What fascination keeps a busy construction engineer at the hobby of butterfly-collecting for a lifetime? One explanation is a service that Colonel Schroeter was able to perform several years ago for Carroll M. Williams, the eminent Harvard University zoologist, who uses insects to study basic life processes. Williams needed a large insect for investigation of metamorphosis and was stymied for lack of supply until he heard about Schroeter and his collection of giant silkworms.

"Our relationship with Colonel Schroeter," writes Williams, "is certainly an excellent illustration of how the amateur can make a distinct contribution to science and share the satisfactions of scientific investigation. The amateur occupies a very special place in entomology because a high proportion of the so-called 'professionals' begin as amateurs. (Later on, incidentally, the complexities of work in a laboratory and an institution may cause them to wish they had remained amateurs!)

"As far as I can judge, Colonel Schroeter was the first to introduce to this country for scientific study and experimentation a wonderful array of 'wild silkworms.' These creatures live in distant parts of the world such as India, Malaya and the slopes of the Himalayas. Colonel Schroeter developed contacts in all these places and has made available to a number of universities and governmental laboratories, including our own, a rich variety of material.

"Certain species of the silkworms have proved strategic for particular types of scientific studies. For example, we have repeatedly called on Colonel Schroeter for specimens of *Antheraea mylitta*, the so-called giant tussah silkworm of India. This exotic creature is one of the world's largest insects, the full-grown caterpillar weighing about 45 grams. It is easy to see how scientists can use beasts of these proportions to answer chemical and physiological questions which would be quite inaccessible in ordinary insects.

"The Colonel has also made available to us considerable information, derived from his own breeding experiments, concerning the care and feeding of these strange species."

About a year ago many newspapers carried a picture of Schroeter with an 11-inch moth which he had reared from an egg the size of a matchhead. This department sought out the Colonel at his home, and it turned out to be a fascinating visit. Colonel Schroeter explained: "The big fellow is an *Attacus edwardsi*. As you can see, its wings are various shades of brown and yellow and contain transparent windows. Specimens caught in the Philippine Islands have a wing span of 14 inches. Larvae of the Atlas species of this moth feed on ailanthus leaves—you know, the tree that grows in Brooklyn. My scientist friends have not shown much interest in the Atlas caterpillar even though it is far from being a pigmy. It is green, finger-sized and has blue horns on its head. Its body looks as if it's frosted with a sugar coating, and natives say that it is delicious.

"I wish you had made your visit a little later in the year. Then I could have shown you a really big moth, *Thysonia aprippina*. It is a native of Brazil. Those I have bred have much larger wing spans than the *Attacus atlas*, which is usually listed as the largest moth in the world. Here is the cocoon of a Thysonia—you can see it is the size of a small sweet potato. You can imagine the proportions of the moth that comes out of it.

"Newspaper reporters make so much fuss over the big fellows that they overlook the really interesting specimens. Take the hybrid luna, for example. Seven years ago an amateur friend of mine in India airmailed to me a dozen cocoons of the Indian moon moth. When the adults emerged some months later, it was evident that they were closely related to the American luna. The two species have about the same shape and size and their wings bear a similar general coloration—a light bluish-green. The wings of the Indian species are distinguished by two patches of red. I decided to try crossing them and finally succeeded last year."

Colonel Schroeter began by acclimatizing the foreign species, which meant breeding several generations of the foreigners here, letting them adapt to local forms of their favorite food plants and to the new environment generally. Then he selected a likely female of the foreign species and mated her with a local male. He has invented a simple gadget to help assure a successful mating. Most amateurs tack a female to a tree by one of her wings and wait for her to attract a mate. "The chief drawback of this technique," says Schroeter, "is that the female's attraction is not limited to mates of her species. When you pin your specimen to a tree you invite predators—other insects, birds and tree toads—to a free dinner. Too often when you come back you find nothing but a pair of wings. Moreover, when you immobilize a single wing the female is apt to thrash around and injure herself. To overcome this difficulty I made what I call a 'mating panel'—a rectangle of Celotex 18 inches long and a foot wide. A screw eye in the center holds a leash of thread, the other end of which is fastened around the female's thorax. With freedom to crawl around on the surface of the panel, she usually quiets down after a few seconds of fluttering. The panel is then suspended by picture wire from the tip of a branch where it is out of the reach of tree toads and free to swing in the breeze. The movement frightens most birds away.

"Moon moths mate at sundown. The next morning the female is transferred to a large paper bag in which she deposits her eggs in two parallel rows. After the eggs have been laid, the bag is cut into little squares, each holding eggs. These are fastened with bits of Scotch tape to the leaves of food plants and surrounded with a bag to prevent the larvae from escaping when the eggs hatch. If you are lucky, the larvae thrive and metamorphosis gets under way. Sometimes the experiment works, but more often it fails. The eggs may be sterile, disease may strike, the food may not be correct. Murphy's law makes no exception of entomology. If anything can go wrong, it will. The failures, however, can be as interesting as the successes, because they pose problems of finding out what went wrong, where, and when. In the case of the moon-luna experiment, nature threw the book at me. But in the end I was rewarded with a beautiful hybrid which bore the characteristics of both parents. Its wing markings fade from bright green into greenish-blue and trail through orange to pink at the wingtips. It is probably the only offspring of this combination in the world."

Colonel Schroeter says most lepidoptera mate readily in captivity. Last season he bred more than 5,000 individuals. Eggs come to him from all parts of the world—sometimes in goose quills and other strange containers. Cocoons arrive in balsa boxes from South America and in bamboo cylinders from the Orient. While there is a law against indiscriminate importation of insects into the United States, the government has issued to Schroeter a special importing license, subject to strict controls.

"Don't let the import restriction on foreign material keep you out of amateur entomology," he urges. "You can collect domestic species to your heart's content without fear of ever exhausting our known varieties. Reference texts and catalogs list them by the thousands, and scores of new descriptions are added each year.

"You will find caterpillars wherever plants grow. The next time you go for a walk, whether in the park, a meadow, or merely in your backyard, take along a paper bag, a piece of string, some note paper, and a pencil. When you find a caterpillar, jot down a short description of it—the color, size, markings, and such other information as you think will help you recognize the creature when you meet another like it. Make a similar record of the plant on which it was feeding. If you already know the name and nature of the plant, so much the better. Be sure to include the date, approximate time of day, and notes on the weather. Then put the bag over the twig or weed on which your specimen is feeding and tie the end closed so it cannot escape. Check up on it a day or so later. You will likely discover that the leaves have been eaten. If so, shift everything to a fresh batch of leaves. You may have to repeat this several times.

"Eventually you will find that your specimen has vanished and a cocoon has taken its place. With luck you may catch the caterpillar in the act of spinning its cocoon. Make full notes of its methods and how long a time it spends in the process. When the cocoon is complete, break off the twig to which it is attached and transfer operations to a small cage, which you can make of window screening. Place the cage outdoors in a location matching as closely as possible that where you found the insect. Some species prefer sunny locations; others do best in shade. After days or weeks—depending upon the species and the season of the year—the adult will emerge, and you will have the thrill of discovering the exotic creature your caterpillar was destined to become."

By starting with the caterpillar instead of with the butterfly, Colonel Schroeter explains, you learn to recognize at first hand three of the four stages in the life cycle of your insect—larva, pupa, and adult. Your notes now give purpose to your future field trips. You hunt for another caterpillar and cocoon of the same species. With luck you may even come across an adult female in the act of laying her eggs. When they have been mounted and labeled, you have the complete life cycle of the insect and the beginning of a collection of scientific value. Although thousands of adult moths and butterflies have been cataloged, the life cycle of a majority of those in nature still awaits description—an ideal project for the amateur who enjoys original work.

"One attractive feature of amateur entomology," says Schroeter, "is the fact that you never run out of interesting projects for your spare hours. Collecting and breeding are merely two facets of the hobby's many sides. For convenience in study, collections must be mounted and labeled. This can be an absorbing pastime the year around. Only the most perfect specimens are selected for mounting. They are killed and stored against the day when bad weather forces you to remain indoors.

"You first stun the insect by pinching the lower side of the thorax lightly between your thumb and index fingers. The thorax is the part of the body, directly back of the head, to which the wings are attached. Stunning is necessary to prevent the insect from fluttering and damaging itself when you drop it into the killing jar. The jar can be any wide-mouthed container with a tight-fitting cover. A layer of absorbent material, such as plaster of Paris, is placed in the bottom and saturated with a tablespoon of Carbona. [I prefer to spray a commercial insecticide, like Raid, on a cotton ball and place it inside the killing jar with the insect. Ed.] The dead specimen is stored in a triangular envelope. The envelopes are numbered to correspond with the entries in your notebook."

In about a week the dead insects become hard and brittle. They must be "relaxed" or softened before mounting. You put the dried insects into a jar containing a rubber sponge or other absorbent moistened with water to which a few drops of carbolic acid have been added. The acid prevents the formation of mold or other microorganisms. A couple of blotters placed between the insects and the sponge will prevent them from getting too wet. After two or three days they are ready for mounting.

Figure 18.1 on page 146 illustrates the details of the procedure. A convenient outfit for mounting consists of a spreading board, two slender strips of glass, tweezers, scissors, pins, and a supply of thin cardboard. A spreading board is easily made from balsa, with a groove in the center for the body of the insect. The slight upward slant of the board on each side of the groove makes allowance for the tendency of the wings to droop as they age. Mounting outfits can be bought from a biological supply house.

To mount the specimen you grasp it by the lower side of the thorax, part the wings by blowing lightly, and thrust a pin through the thorax from the top. The pin should be inserted into the insect just far enough to bring the point of wing attachment level with the surface of the spreading board when the pin has been forced into the bottom of the groove. Then blow the wings apart again and place the specimen on the board, weighting down the wings with strips of glass. Each glass is lifted in turn just enough to

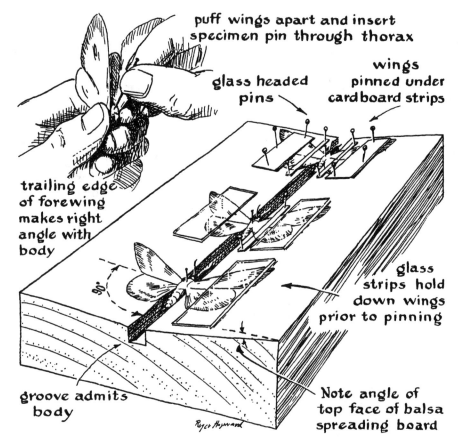

Figure 18.1   Steps in the mounting of specimens for display.

permit pulling the forewings forward by means of a pin inserted behind one of the heavy veins. When the trailing edge of the forewing makes a right angle with the axis of the body, it is pinned down with strips of cardboard as shown. Wider strips of cardboard are then pinned in place of the glass weights. The specimens will be dry enough in about a week for transfer to the display case.

Eggs and pupae are mounted without special preparation. Cement the eggs to paper strips of contrasting color. Pin the pupa as though it were a dried adult. A larva must be degutted and inflated before mounting. After killing, place the larva on a square of blotting paper and, with a fine scalpel, enlarge the anal orifice slightly. Then, beginning at the head, squeeze the viscera out by rolling a pencil down the body. The carcass is restored to normal shape by inflating it with a syringe, which you can make yourself. Heat a section of quarter-inch glass tubing to a dull red and quickly draw one end into a fine nozzle, somewhat thinner than the small end of a medicine dropper. Fit the large end with a rubber bulb. You then inflate the larva by inserting the nozzle into its anal opening and squeezing the bulb sharply. The anal opening is closed with a bit of Scotch tape until the tissues harden.

For convenience in subsequent study, specimens are generally pinned to the bottom of a glass-topped display tray. Many arrange the eggs, larva, pupa, and adult of each species as a group.

Colonel Schroeter recently gave his entire collection of thousands of specimens to the University of Connecticut. J. A. Manter, the university's zoologist, writes: "The Schroeter collection is the most colorful that I have ever seen. Every division of world macrolepidoptera is represented by rare specimens, and it is especially remarkable because of their excellent condition. Such a collection is often spoken of as an 'Oh, my!' one—the reaction it evokes from visitors as the trays are successively pulled into view. Colonel Schroeter's devotion to amateur entomology has resulted in a lasting contribution to science from which future generations of students will derive both knowledge and enjoyment."

The late William Morton Wheeler, the great Harvard entomologist, once summed up the joys of the amateur in these words: "We should realize, like the amateur, that the organic world is also an inexhaustible source of spiritual and aesthetic delight. Especially in college we are unfaithful to our trust if we allow biology to become a colorless, aridly scientific discipline devoid of living contact with the humanities. We should all be happier if we were less completely obsessed by problems and somewhat more accessible to the aesthetic and emotional appeal of our materials. It is doubtful whether, in the end, the growth of biological science would be appreciably retarded. It quite saddens me to think that when I cross the Styx, I may find myself among so many professional biologists, condemned to keep on trying to solve problems, and that Pluto, or whoever is in charge down there now, may condemn me to sit forever trying to identify specimens from my own diagnoses while amateur entomologists, who have not been damned professors, are permitted to roam at will among the fragrant Elysian meadows netting gorgeous, ghostly butterflies until the end of time."

I've been fascinated by butterflies and skippers ever since I was a boy. And I've bred them off and on throughout my adult life. In fact, when I was in my mid-twenties, I used to keep my trusty homemade net at the ready just in case a particularly interesting specimen happened into the planted courtyard behind my apartment. Once, while stepping out of a midday shower I saw a mating pair of swallowtails out my bathroom window tantalizingly close to the back door. Realizing that I had only seconds to act I grabbed my net and leaped into action *au naturel*, startling the butterflies (which I quickly captured) as well as a few sunbathing neighbors.

So it was inevitable that I would eventually write an article about raising butterflies. Until then "The Amateur Scientist" had only treated the subject once, some 43 years earlier. You've just read that article; my article follows. Although there is some overlap, they provide a great deal of complementary information. I hope you'll enjoy this delightful pastime as much as I have. Ed.

Some of my earliest adventures in science came while I was a small boy hunting insects in my mother's flower garden. She loved her garden for its explosions of yellow, lavender and violet blossoms, but I was much more interested in stalking the ostentatious visitors they attracted. My mother's flowers served as sugary ports of call for butterflies so dazzling that spring still finds me netting and occasionally raising these delightful insects.

Lepidoptera (a term that includes butterflies, moths, and skippers) is perhaps the most widely studied order of insects. Yet with only modest equipment, the amateur lepidopterist can find almost endless diversion and even do original research. Populations can respond dramatically to changing habitats, and each specimen tells a story.

The first requirement for a butterfly hunter is, of course, a deep gauze net. You can buy one from a biologists' supply house, but diehard do-it-yourselfers may want to make their own *[see sidebar on page 150]*.

Never chase a flying insect. You'll run yourself into the dirt in half an hour with nothing to show for your exhaustion. Instead approach your quarry slowly while it is feeding and scoop it up from behind. You may also be able to catch butterflies from a stationary position as they flutter past. Always let the net overtake them rather than swooping it toward them head-on. Make sure to jerk the net closed with a quick 90-degree twist at the end of your swing to keep your prey from escaping.

If you want to assemble a butterfly collection, you should raise your own specimens—they will be free of the parasites that can mar wild exemplars, and they are fascinating to study. Luckily, butterflies are easy to rear. The sex of members of many species can be determined by the patterns of spots on their wings. If you are uncertain, catch at least four members of the species you wish to breed. You will then have seven chances in eight of having at least one mating pair. Place them in a glass terrarium full of the butterfly's food plants. (Field observations will help you here, or *see the Further Resources list on page 151* for suggestions.) It's okay to use cuttings, but replace them often to keep them fresh. Affix cotton netting or muslin over the terrarium's top and place it outside and out of direct sunlight.

A female can also lure her own males for mating. But please don't, as some entomologists suggest, tack one of the female's wings to a tree to immobilize her. Her agonized fluttering will very likely injure her severely and attract more predators than suitors. It is much better to create a short leash by gently knotting a 10-centimeter thread between her thorax and abdomen and tacking the end to the center of a square of thick poster board about 20 centimeters on a side. Secure a bouquet of her favorite flowers to the square and hang the assembly horizontally in the late afternoon near where you caught her. This platform will separate her from tree-dwelling predators, and its swaying will scare away most birds.

In the early morning, transfer your hopeful mother to a net-lined enclosure surrounding a plant on which her species' caterpillars can feed. Caterpillars can be picky, so some butterflies will lay eggs only on particular kinds of plants. If you don't know which plant to use, secure cuttings from several kinds near where the female was caught. The eggs might be sterile, but if they hatch, keep track of which caterpillars do best.

Plastic two-liter soda containers make ideal caterpillar rookeries. Cut off the top and tape a layer of muslin over it, then place the eggs and your caterpillar's favorite food plant (or cuttings from the plant the female laid her eggs on) inside. A smidgen of petroleum jelly on the inner lip will keep the hatchlings from climbing up the side and fouling themselves in the netting.

Keep your caterpillars outdoors with sunlight and temperature as close as possible to their normal habitat. As your new family matures, you'll need more rookeries. A caterpillar must get into the right position to pupate, and more than about five siblings often means too few places in a small container. Arrange twigs in the containers to provide suitable niches. If you rear too many caterpillars for your needs, release the excess creatures on live food plants in their native environment. Transfer any chrysalides or cocoons that develop to your outdoor terrarium and record their daily progress in your field notebook. If all goes well, they will ultimately become magnificent adults.

Select only the best specimens for mounting. Traditionally, two are chosen—one mounted face down and the other face up to display both sides of the wings. Place cotton balls soaked in either a commercial insecticide or fingernail polish remover into a tea strainer. Hang the strainer

Constructing a Butterfly Net

The scoop must be fashioned from soft fine-mesh cotton netting and be at least two feet deep. Use a sewing machine to embed a length of twine along the edge that will become the side seam; this reinforcement will keep the seam from unraveling.

Stitch fabric reinforcement onto the edge that will become the opening and then sew the net and fabric around a length of coat hanger wire. Bend the wire into a circle around the bottom of a stew pot, using pliers to bend the ends outward.

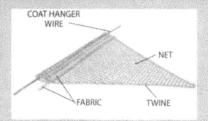
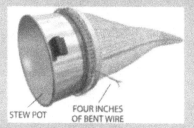

Close the side seam using a hand needle to spiral the thread around the twine. (Carpet thread creates a virtually indestructible seal.) After fashioning the net, make two four-inch-deep crossing cuts with a coping saw along a broomstick's axis.

Insert the ends of the net's wire frame and pour a generous helping of 24-hour epoxy into the slots. Finally, tightly wrap the assembly with a layer of cotton clothesline. My first homemade net survived more than 10 years of vigorous use.

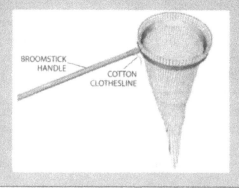

with a thread near the top of an airtight jar and place a few layers of paper towel on the jar's bottom. Stun the insect with a firm (yet careful) pinch on the underside of its thorax, drop it into the jar and secure the lid. Return in about 30 minutes to remove your specimen from the jar.

If you can't mount the specimen immediately, store it in a triangularly folded piece of blotting paper. Keep stored butterflies in a sealed container in the freezer until you can mount them. Make sure the specimens have returned to room temperature before you proceed.

Figure 18.1 on page 146 shows the procedure for mounting butterflies. Cork or balsa wood make good surfaces for pinning. These should be slightly angled as shown, to compensate for the wings' tendency to droop. Begin by inserting an insect pin through the right dorsal side of the thorax. (Never substitute sewing pins for insect pins; they are too thick and will rust.) Spread and secure the wings and then attach the insect to the center of the mounting board. Keep the abdomen from sagging by crossing two pins directly underneath it. Let the specimen dry out for at least a week before transferring it to a permanent case.

Look out for minuscule mounds of dust beneath your specimens that tip off the activity of tiny insect pests that may be slowly devouring your collection. Make sure each case is tightly sealed with a few mothballs inside to control these marauders.

Serious collectors should connect with their nearest natural history museum. Many museums continually collect local species to monitor the changes in these populations. Collecting for an institution is a great way to advance science and give conservationists the raw data they need to help protect the environment.

Further Resources

Books:

The Amateur Naturalist. Gerald Durrell. Alfred Knopf, 1982.

Butterflies East of the Great Plains. Paul A. Opler and George O. Krizek. Johns Hopkins University Press, 1984.

The Butterflies of North America. James A. Scott. Stanford University Press, 1986.

The Practical Entomologist. Rick Imes. Simon & Schuster, 1992.

Biological Supplies:

BioQuip Products. 17803 LaSalle Ave., Gardena, CA 90248. Telephone: 310-324-0620. E-mail: *bioquip@aol.com*

Butterfly World. Tradewinds Park, 3600 West Sample Rd., Coconut Creek, FL 33073. Telephone: 954-977-4400.

Organizations for Butterfly Enthusiasts:

Sonoran Arthropod Studies Institute. P.O. Box 5624, Tucson, AZ 85703-0624. Telephone: 520-883-3945. E-mail: *arthrostud@aol.com*

Young Entomologists' Society. Gary and Dianna Dunn, 1915 Peggy Place, Lansing, MI 48910-2553. Telephone: 517-887-0499. E-mail: *yesbug@aol.com*

Lepidoptera Research Foundation. 9620 Heather Rd., Beverly Hills, CA 90210. Telephone: 310-274-1052. E-mail: *Mattoni@ucla.edu*

Lepidopterists' Society. Membership: 1900 John St., Manhattan Beach, CA 90266-2608. Telephone: 310-545-9415.

19 MEASURING INSECT METABOLISM

by Shawn Carlson, December 1995

etabolism is basic to life. Everything that breathes combines chemical energy stored in its tissues with oxygen contained in the atmosphere to liberate energy to grow, move, and reproduce. Despite long-running efforts, biologists have so far surveyed only the basics of metabolism. Life is so diverse that discoveries await anyone who ventures carefully into these deep waters. The earth harbors perhaps as many as 10 million species of insects, and very few of these have been studied in any detail. So insects offer myriad opportunities for amateur exploration.

It is actually quite simple to measure the metabolism of an insect. When an organism is enclosed in an air-tight container, its respiration removes oxygen molecules from the air and releases carbon dioxide. Often fewer molecules are added to the air than are removed. The resulting loss causes the pressure inside the container to fall.

That pressure drop is key to measuring metabolism, and it can easily be observed using a device called a Warburg apparatus *[see Figure 19.1 on page 154]*. The instrument consists of two stoppered test tubes connected by a capillary tube. A tiny droplet of oil (or soap) in the capillary tube moves in response to pressure differences between the test tubes. Therefore, as the respiration of an insect in one of the test tubes causes a decrease in pressure in that tube, the oil drop will slip toward it. You can witness this movement if you warm one test tube with your hand. That makes the air inside expand and push the droplet to the cooler test tube. To quantify the movement of the droplet, photocopy a ruler with millimeter gradations and tape the copy to the capillary tube.

You want the drop itself to offer the smallest possible resistence to the force generated by the change in pressure so make sure that the drop is as

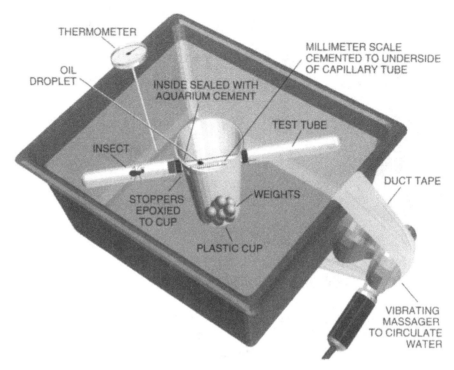

Figure 19.1 Insect respiration can be calculated by knowing the air pressure, the water temperature, and the distance covered by an oil droplet in the capillary tube.

small as you can possibly make it. You also want to use a liquid with the lowest possible viscosity so that it offers the least frictional resistance inside the tube. You can use water or vegetable oil stained with food coloring (to make it more visible), or motor oil. I've also had good results with WD-40. Also, you might also enjoy experimenting with brake fluid which is colored and which, like oil, does not evaporate. And brake fluid also has a low viscosity.

Getting the droplet into the center of the tube can be tricky. First, thread the capillary tube through the rubber stoppers. Then dip the capillary tube into the fluid. Suck the fluid about ¼ of the way up the tube. Place a test tube over the end you just dipped and gently grasp the test tube allowing heat from your hand to warm the air. As the air warms, it expands and will push the fluid up into the capillary tube.

Completely coat the inside of the capillary tube by letting the fluid run all the way to the other side of the capillary tube. By adjusting your grip on the test tube you can control how fast the plug moves. When it reaches the end, grip it very lightly so the plug moves very slowly. You want the plug to be just a few millimeters across so drive most of the fluid out through the end, while dabbing the end with a paper towel to catch it. When you've got about five millimeters of fluid left in the capillary tube, remove your hand and plunge the test tube into a room temperature water bath. Gently place a test tube over the other stopper. Warm this with your hand and the plug will move away from this end. When the plug is positioned in the center of the capillary tube remove the test tubes. Make sure you move the plug slowly. If you try to position it too fast it will splinter into a number of fragments and you'll have to start all over again.

It's a bit tricky to get this right so you'll have to practice. Because of thermal inertia, the plug will move a bit even after you take your hand off the test tube and so you'll have to learn how to anticipate its motions. It took me about an hour to get the hang of it so don't get discouraged.

As the warming with your hand demonstrates, the device is sensitive to small temperature differences between the test tubes. The best way to ensure equal temperatures is to submerge the tubes in a large basin of water. To keep them below the surface, attach them to the sides of the plastic cup. Weigh the cup down with sand, pebbles, or a fistful of spare change. The cup also permits you to view the oil in the capillary tube in dry air. To reduce further the effects of temperature gradients, set the water moving slowly with a handheld body massager.

If you know the air pressure, the water temperature and the distance the oil drop moves, you can determine the number of molecules the insect respires. The box lists the exact relations.

The next step is to find the ratio of carbon dioxide produced to oxygen consumed. Called the respiratory quotient, it represents a fundamental measure of metabolism. It tells you what biological fuel the organism is burning. If it is converting sugar, the ratio is 1; for fat, about 0.70; for protein, about 0.80; for alcohol, about 0.67. For most creatures, the quotient ranges from 0.72 to 0.97, because organisms metabolize several kinds of energy sources simultaneously.

To measure the respiratory quotient, you will need some sodium hydroxide (NaOH), which absorbs carbon dioxide from the air. Purchase this compound in solid form from any chemical supply house; check your Yellow Pages. Alternately, you can purchase this and other chemical supplies from the Society for Amateur Scientists *[see page 159]*. But watch out—sodium hydroxide is caustic and will burn skin and eyes if not handled properly. Rubber gloves and safety goggles should be worn.

Before conducting trials, you will need to clear all the carbon dioxide out of the test tubes. Place several grams of NaOH in just one test tube. The toe of a nylon stocking makes an excellent pouch to hold the chemical; ball

Calculating Respiration

The number of molecules the organism respires, or ΔN, is equal to $9.655 \times 10^{16} \, PA\Delta L/T$. Here P is the atmospheric pressure in centimeters of mercury (if you do not have a barometer, call your local weather service whenever you conduct a trial), A is the cross-sectional area of the inside of the capillary tube in square millimeters, ΔL is the distance the droplet moves in millimeters, and T is the temperature of the water bath in kelvins. To convert Celsius to kelvins, add 273.15 degrees. The numerical constant is my own derivation from the physics involved.

Figuring the respiratory quotient—the ratio of carbon dioxide molecules released to the number of oxygen molecules consumed—is not much more difficult. It equals

$$\frac{\Delta N_{O+NaOH} - \Delta N_O}{\Delta N_{O+NaOH} + \Delta N_O} = \frac{\Delta L_{O+NaOH} - \Delta L_O}{\Delta L_{O+NaOH} + \Delta L_O}$$

where ΔN_O is the number of molecules removed by just the organism; and ΔN_{O+NaOH} is the number of molecules removed when both the organism and the NaOH are in the test tube.

As is the case for ΔN, the subscripts denote the conditions of the trial: either with the organism alone or with the organism and the NaOH. Note that you do not need to know the atmospheric pressure, temperature or area of the capillary tube if you are looking solely for the respiratory quotient.

If you use the differential pressure transducer, the equations are slightly different. The value of ΔN equals $1.804 \times 10^{19} \, V\Delta P/T$, where V is the volume in cubic centimeters of the test tube containing the organism (allow for the volume taken up by the insect and the NaOH), ΔP is the pressure change in inches of water, and T is the temperature of the water bath in kelvins. The respiratory quotient then equals $(\Delta P_{O+NaOH} - \Delta P_O)/(\Delta P_{O+NaOH} + \Delta P_O)$, where ΔP_{O+NaOH} is the pressure change measured with both the organism and the NaOH in the test tube, and ΔP_O is the pressure change measured when the organism is by itself.

the nylon up on the open side to prevent the insect from touching the NaOH. Measure how long it takes for the droplet to stop moving, that is, for the NaOH to remove the carbon dioxide from the air. Make sure to wait at least that long before you start each trial. You will want the system to come to equilibrium quickly, so use a lot of NaOH. (Alert readers may wonder about water vapor, which NaOH also absorbs; for technical reasons, it will not affect the measurements.)

Now you are ready to begin your experiments. First, place both the wrapped NaOH and the creature inside one test tube and just NaOH in the other. Measure how long it takes for the droplet to move at least five times the smallest distance marked on your scale. Then run a second trial for exactly the same length of time, but with only the organism. The box lists the equations needed to obtain the respiratory quotient. Note that your results will be valid only if the organism is in the same physical state in both trials (not calm in one and agitated in the other, for example), if the NaOH removes all the carbon dioxide from the air before that trial begins and if both trials are run for exactly the same length of time.

This method, though time-honored and quite effective, is tedious and requires a great deal of experience and skill to obtain consistent results. However, with an investment of about $100, you can easily collect professional-quality data suitable for publication in a research journal. You will need to buy an electronic differential pressure transducer, a device that converts pressure differences into voltages that can be measured with a voltmeter. I used a Honeywell model (No. 163PCO1D36) that registers pressure differences as small as 0.0003 percent of one atmosphere. (For more information about this sensor, call the Honeywell Corporation at 1-800-537-6945.)

The power-supply circuit for the device could not be simpler. It consists of an AC-to-DC adapter wired to a type 7812 integrated-circuit chip. Its voltage drifts a bit, causing the transducer's output to wander about 10 millivolts, but that should not pose much of a problem. You will need to calibrate the transducer with a manometer—a clear, U-shaped plastic tube with some water inside *[see Figure 19.2]*. I made mine out of two thin, stiff tubes I found in the garden department of a hardware store. Aquarium pet stores also stock similar kinds of tubes. I joined the pieces by inserting each into opposite ends of a

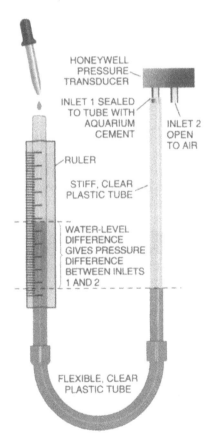

HONEYWELL
PRESSURE
TRANSDUCER

INLET 1 SEALED
TO TUBE WITH
AQUARIUM
CEMENT

INLET 2
OPEN
TO AIR

RULER

STIFF, CLEAR
PLASTIC TUBE

WATER-LEVEL
DIFFERENCE
GIVES PRESSURE
DIFFERENCE
BETWEEN INLETS
1 AND 2

FLEXIBLE, CLEAR
PLASTIC TUBE

Figure 19.2 Calibration of the Honeywell transducer relies on a difference of water level in the two stiff tubes.

A Few Important Tips

Keep lots of spare capillary tubes around. No matter how careful you are, you are going to break them. Try a number of different diameters of capillary tubes. Experiment. Different tubes will be suited for different conditions. It will take some practical hands-on experience to select the best tube size for your application. Shore up the inside of the plastic cup with a wooden disk that fits just below the capillary tube. Paint the disk black or cover it with felt. When you plunge the plastic cup into the water bath the water pressure will distort the cup a little and will place stress on the capillary tube. The disk reduces this stress and helps protect it against breakage. It also provides a dark background against which the plug can be clearly seen. (I made this modification after the article was in press, so the disk doesn't appear in the figure.)

Place both test tubes on the stoppers very carefully and at the same time. When the test tube seals over the stopper, the pressure inside will go up because the volume decreases slightly as the test tube is pushed tight. That change in pressure can be enough to propel the plug right out the other side. You can balance this by placing both test tubes on slowly and at the same time. In fact, with a little practice, you can do it without moving the plug at all. To remove a test tube from the stopper, pinch the stopper firmly to hold it in place and gently twist the test tube as you pull it off. By holding the stopper with your fingers you are reducing the stress felt by the capillary tube. Tie a small fish weight at the end of the test tubes to compensate for the buoyant force experienced by the test tubes in the water bath. This will also reduce the stress on the capillary tube.

The buoyant force is equal to the volume of the test tube in cubic meters times the density of water in kilograms per cubic meter times g (the acceleration due to gravity = 9.8 meters per square second). This force acts at the center of the test tube and produces a torque that tries to rotate the test tube upward and break the test tube off at the capillary tube.

To compensate, calculate the mass of the water displaced. It's the density of water (1 gram per cubic centimeter) times the number of cubic centimeters of the test tube. Find two fish weights that are about half this weight and attach each to a loop of thread. When you're lowering the test tubes into the water bath, hook the weights over the ends of the test tubes. Since these weights are about twice as far away from the cup, they only need to be half as heavy as the buoyant force to cancel the torque which would otherwise try to snap the capillary tubes. Ed.

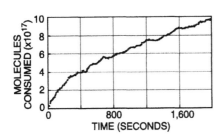

Figure 19.3 Beetle breaths recorded over time indicate, by dips in the plotted curve, that the insect "exhales" about once every seven minutes. (Data were taken without NaOH.)

six-inch-long flexible acrylic tube. The difference in height in the water column on either side of the U is a direct measure of the pressure. You can plot the output voltage against that difference, in inches (to match the standard units that pressure transducers made in the U.S. use).

The transducer has two inlets that permit easy connection to the rubber stoppers of each test tube in the Warburg apparatus. I used this setup to measure the respiratory quotient of a beetle, which I found to be 0.701, averaged over several breaths *[see Figure 19.3]*. Of course, you can also measure the metabolism of other living things: mushrooms, seeds, bread mold, to name a few.

For more information about this and other amateur science projects, check out the Society for Amateur Scientist's World Wide Web page at *www.sas.org*. You can call them at 401-823-7800, or write them at the Society for Amateur Scientists, 5600 Post Road, #114-341, East Greenwich, RI 02818.

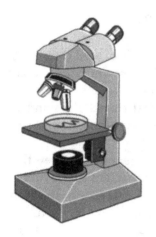

20 GETTING INSIDE AN ANT'S HEAD

by Shawn Carlson, June 1997

The toy microscope set I got one childhood Christmas almost put me off microscopy for good. Squinting through a cheap plastic eyepiece, manipulating the sloppy focus control with one hand, and struggling with the other to position the factory-prepared specimen slides resulted in more neck aches and frustration than reward. So years later, when a friend offered me his old professional-quality microscope for a mere fraction of its market value, I hesitated. Besides, I could scarcely see through the dirt-encrusted optics, and the gearing for the traveling stage was so gunked up that it scarcely traveled at all. But a little patience, along with cleaning fluid and grease for the gears, quickly restored this binocular beauty to mint condition. Today I'm convinced it was the best $100 I ever spent. My vintage Spencer has been my constant companion for more than a decade. With it I have poked about inside plant cells, swum abreast of wriggling sperm, and gotten delightfully lost inside floating forests of phytoplankton. Unfortunately, no biological specimen can be enjoyed under a microscope for long before natural processes begin to destroy it. Tissues quickly dry out and shrink, and bacteria begin breaking down the delicate structures of life almost immediately after a specimen dies.

Microscopists therefore routinely preserve their treasures before scrutinizing them carefully. The process, called fixation, replaces the water in the specimen's tissues with a chemical that is incompatible with life. The fluid pressure keeps the cells plump while the chemical kills decay-producing bacteria. Cellular proteins are not soluble in many fixing agents, however; they precipitate to form a sort of plaster that sticks to cell walls. This process stiffens the cell wall and can actually increase its index of refraction, making the cell stand out under the microscope, but it ultimately destroys potentially useful information.

When amateur scientist extraordinaire George Schmermund of Vista, California, recently brought together his interests in insects and microscopy, he uncovered a wonderful fact. Different tissues within an insect absorb the fixing agent he used at different stages in the fixing process. By observing the specimen throughout fixation, rather than waiting until the process is complete, as microscopists are often taught to do, he found that separate organ systems within the animal became highlighted sequentially. This let Schmermund take the stunning pictures of insect insides that you see here *[Figures 20.1 and 20.2]*. His technique empowers any amateur

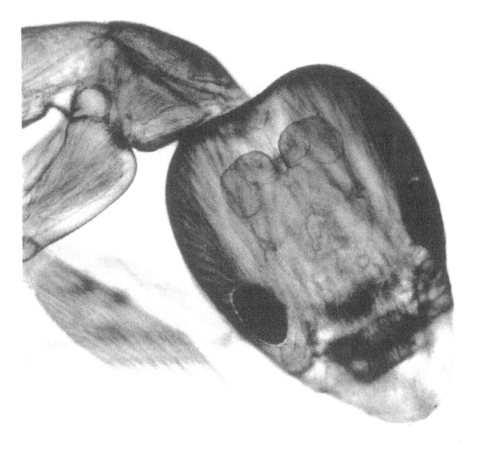

Figure 20.1 This ant's head, photographed during fixation, shows details that later disappear.

Figure 20.2 The flea leg shown here (top) had clearly visible muscle striations. The mouthparts of a wood tick are revealed in detail (bottom).

microscopist to explore insect interiors in a detail unmatched by any other method I know.

Begin your own fantastic voyage by securing some suitable specimens. Small ants and fleas make ideal subjects. Schmermund kills them by placing them in his freezer for a few hours. His fixing agent is ordinary isopropyl (rubbing) alcohol. Isopropyl alcohol is inexpensive, readily available, and easy to handle.

My local drugstore stocks isopropyl alcohol in concentrations of 70 and 91 percent. Buy the highest concentration you can find. For reasons that will become clear in a moment, if you intend to create a permanent library of mounted specimens you'll need to purchase pure (anhydrous) isopropyl alcohol from a chemical supply company. (Note that denatured alcohol is a blend of mostly ethanol and methanol; it is not isopropyl alcohol. Both substances are also fixing agents, but methanol is highly poisonous. If you choose to experiment with denatured alcohol, take appropriate precautions to protect your family and pets.)

The box gives a recipe for diluting your fixing solution to any desired concentration. Schmermund took these photographs using only two dilutions: 35 and 70 percent. The time required at each stage depends on the specimen's size. Soak garden ants and fleas for at least one hour. Larger insects may take as long as six hours. The volume of solution should be at least 20 times that of your specimen's body, but for ant-size specimens that is still a tiny amount of alcohol. Schmermund soaks his specimens in bottle caps and transfers the chemicals with eyedroppers.

To get views like those you see here, you must regularly check the insect while fixing. Transfer the insect to a well slide (a slide with a polished depression to receive the sample) with a drop of fixing solution, rest

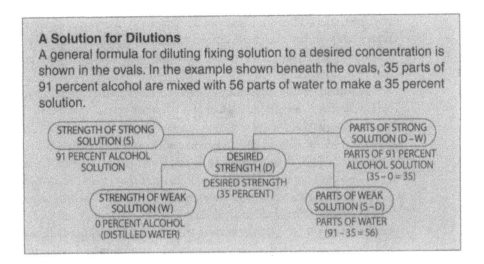

A Solution for Dilutions
A general formula for diluting fixing solution to a desired concentration is shown in the ovals. In the example shown beneath the ovals, 35 parts of 91 percent alcohol are mixed with 56 parts of water to make a 35 percent solution.

a slide cover on top, then set sail into uncharted waters. Transfer and position the specimen with laboratory tweezers (available at a laboratory supply store and often at your local swap meet) and an eyedropper.

Unfortunately, neither water nor alcohol mixes with the compounds used to glue your specimen permanently to the slide. So to share your specimen with posterity, you must first replace every trace of both water and alcohol with a solvent that will mix with the glue. The process that removes the alcohol is called clearing. Clearing will make your specimen largely transparent and will destroy much of the contrast created during the fixing stages. You can lock in some permanent contrast by staining your specimen with various commercially available dyes, such as borax carmine or alum cochineal. But no method I know of produces the level of contrast in specific tissues that naturally develops during the fixing process.

Xylene is the clearing agent of choice for the amateur scientist. I recently purchased 32 ounces (a lifetime supply!) for $5 in the paint section of my local hardware store. But watch out: xylene is poisonous, it dissolves plastic on contact, and its fumes can make you quite sick. So be particularly careful to protect yourself and your family and pets. In addition, although water mixes with alcohol, and alcohol mixes with xylene, water and xylene do not mix. The water must be completely removed with a final fixing step in 100 percent anhydrous isopropyl alcohol before clearing.

Once the specimen has been soaked in pure alcohol for two hours or more, place it in a solution of equal parts of xylene and anhydrous isopropyl alcohol. Cover your container to keep the xylene from evaporating. Your specimen will probably float at first. Let it soak for an hour or two after it sinks to the bottom and then replace the mixture with pure xylene. Your specimen should become fairly translucent at this stage. Dark patches mark water that was not removed during fixation. These can often be cleared up by returning the specimen to the anhydrous alcohol for several hours and then reclearing it.

Scientific supply companies sell Canada Balsam and a variety of plastic resins for the permanent mounting of specimens. The resin dissolves in xylene, and so it infuses into the insect's tissues before hardening. To make your final slide, place your cleared specimen in the depression in the well slide, add a drop of resin, and gently angle the slide cover down on top, being careful not to trap any air bubbles. Let the resin set before moving the slide.

One final note. Some microscopists have marveled at the striking depth of field visible in Schmermund's handiwork. Here's his secret. These photographs were taken with 35-millimeter film at low magnification and then enlarged in a darkroom. Low magnification means good depth of field.

As a service to amateur scientists everywhere, the Society for Amateur Scientists (SAS) has put together a kit that contains everything you'll need to fix, clear, and mount small insects. The kit contains anhydrous isopropyl alcohol, xylene, laboratory tweezers, six well slides, a small killing jar for your freezer, small petri dishes for fixing and clearing, slide covers, and mounting resin. Send $54.95 plus $5 shipping to the Society for Amateur Scientists, 5600 Post Road, #114-341, East Greenwich, RI 02818. For telephone orders, call 1-401-823-7800. For information about this and other amateur science projects, check out the SAS's World Wide Web site at *www.sas.org* or call 1-401-823-7800.

21 DETECTING INSECT HEARTBEATS

by Shawn Carlson, August 1996

There's a lot going on down among the microns. What we perceive as a rigid surface squashes easily under a finger's gentle pressure when viewed from a distance of a millionth of a meter. Increasing the temperature sends objects at that scale into even more violent upheavals. Biological processes reshape many living things on this scale. For example, every beat of an insect's dorsal vessel—essentially, its heart—flexes its abdomen by a few microns.

Now, thanks to John R. B. Lighton, a biologist at the University of Nevada, these tiny movements can be readily detected. (Lighton is not only a world-renowned physiologist but also a kindred spirit to amateur scientists everywhere, always striving to find the most direct and least expensive solution to vexing experimental challenges.) He realized that by detecting the microscopic flexings of an insect's body, he could in effect put a tiny stethoscope on the creature. Lighton's ingenious method allows experimenters to embark on a fantastic voyage into the microscopic universe. Now anyone can detect movements as small as half a micron—about the wavelength of visible light—for less than $40.

Lighton senses micromotions by using minuscule magnets that he attaches to the moving objects. He then relies on a special sensor that picks up the variations in the magnetic field caused by the shifting magnet.

The sensitivity of Lighton's detector depends on the fact that all magnets are dipolar; they have a north pole on one end and a south pole on the other. These poles would cancel each other perfectly if they weren't separated by the length of the magnet. This self-cancellation quality makes the strength of a magnetic field fall quite fast over space. Tripling the range to the magnet weakens the field by a factor of 27—the cube of the distance. The size of the magnet sets the scale by which this falloff can be quantified. The closer together the magnetic poles are (that is, the smaller the magnet), the more rapidly the magnetic field changes over distance. That in turn produces a larger signal for a micron-size shift.

It's easy to get micromagnets. You can buy so-called rare-earth magnets from Radio Shack (part number 64-1895), which sell for less than $2 a pair. They are tiny disks about 0.48 centimeter in diameter and 0.16 centimeter high (³⁄₁₆ by ¹⁄₁₆ inch). At the surface, the magnetic field, which is oriented perpendicular to the flat part of the disk, is about 20,000 times that of the earth.

Even these tiny magnets are too big for an insect stethoscope, so crush one with a pair of pliers. Made from a brittle ceramic, they will shatter into little shards. You need to make sure, though, that you know the direction in which the magnetic field of these shards points. Using nonmagnetic tweezers, position a fragment on a piece of wax paper. Placing the second magnet underneath the paper forces the fragment to align with the bigger field. Then dab a dollop of paint or five-minute epoxy over the magnetic speck. Once it sets, the magnetic fragment will easily peel off the paper. Make at least 10 of these magnetic chunks, all slightly different in size, to see which one works best for your application *[see Figure 21.1]*.

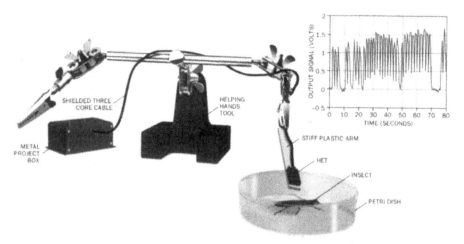

Figure 21.1 An insect stethoscope relies on a chip called a Hall effect transducer (HET), which is held to within a centimeter of the subject by a 12-centimeter-long plastic arm. The device recorded the "heartbeats" of a cockroach nymph (graph). The heart briefly paused after the 70-second mark because the experimenter distracted the nymph with a hand wave.

A Hall effect transducer (HET) senses the changes in a magnetic field. A modern-day silicon miracle, it is small, extremely sensitive, and easy to use. Lighton recommends model SS94A1F from Honeywell Micro Switch in Freeport, Ill.; call 800-537-6945 for a distributor. A bargain at less than $20, this device changes its output by 25 millivolts for each one-gauss shift in a magnetic field. (But be advised, Honeywell's minimum order on this part is five units. However, the Society for Amateur Scientists is making this part available in single quantities. Contact them at SAS, 5600 Post Road, #114-341, East Greenwich, RI 02818; 401-823-7800 or *www.sas.org.*)

Secure your HET no more than one centimeter away from your subject. For instance, if you are monitoring insects, you can epoxy the HET to a rigid piece of plastic and hold it above the subject with a device called Helping Hands, a soldering aid sold by Radio Shack.

A HET records all magnetic fields, including the earth's. This indiscriminateness means that the detector will always produce a large constant voltage signal (created by the earth and the magnet). On top of this voltage constant will be the small changing signal created by the magnet's motion.

Most op-amps aren't very good at detecting a small changing signal that rides atop a large constant one. When faced with such a challenge many experienced experimenters rely on special devices called instrumentation amplifiers. Like op-amps, instrumentation amplifiers are available as inexpensive, integrated circuits. Entry-level devices cost about $5; the Cadillacs of these chips sell for about $20. The AD524 from Analog Devices in Norwood, Massachusetts, works well; to order, call 800-262-5643, extension 3. You can also construct an instrumentation amplifier from three type 741 op-amps, but you'll probably have fewer headaches if you go with a prefab unit.

If you ever want to use this technique to monitor temperature or some other signal that varies slowly, use Lighton's slowly varying signal rendition of the circuit. For flexing insect abdomens and other activities that change significantly over 30 seconds or so, use the quickly varying signal circuit. The circuit employs a clever technique that should be in every amateur's (and professional's) tool kit. The method splits in two the voltage from the HET. One signal goes into the amplifier's positive input. The other goes into a low-pass filter that only passes signals that oscillate slower than about one cycle every 30 seconds. Because an insect's heart contracts in much less time, the filter strips out that signal and passes the large constant voltage (the DC offset). This filtered voltage is then fed into the instrumentation amplifier's negative input. An instrumentation amplifier boosts the difference between its two inputs, so the troublesome offset voltage is automatically subtracted, leaving only the coveted signal *[see Figure 21.2]*.

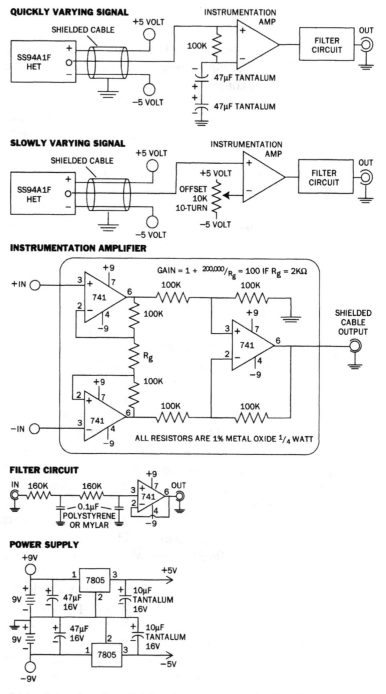

QUICKLY VARYING SIGNAL

INSTRUMENTATION AMP

SHIELDED CABLE

+5 VOLT

SS94A1F HET

100K

FILTER CIRCUIT

OUT

47µF TANTALUM

47µF TANTALUM

−5 VOLT

SLOWLY VARYING SIGNAL

INSTRUMENTATION AMP

SHIELDED CABLE

+5 VOLT

SS94A1F HET

+5 VOLT

OFFSET 10K 10-TURN

FILTER CIRCUIT

OUT

−5 VOLT

−5 VOLT

INSTRUMENTATION AMPLIFIER

+IN

+9

GAIN = 1 + 200,000/R_g = 100 IF R_g = 2KΩ

741

100K

100K

100K

+9

SHIELDED CABLE OUTPUT

R_g

741

−9

+9

100K

741

100K

100K

−IN

741

−9

ALL RESISTORS ARE 1% METAL OXIDE ¹/₄ WATT

FILTER CIRCUIT

IN 160K 160K

+9

OUT

741

0.1µF POLYSTYRENE OR MYLAR

−9

POWER SUPPLY

+9V

1 7805 3

+5V

9V

47µF 16V

10µF TANTALUM 16V

9V

47µF 16V

10µF TANTALUM 16V

1 7805 3

−5V

−9V

Figure 21.2 Rate of motion dictates the necessary circuitry. If the signal changes much over about 30 seconds, choose the quickly varying circuit. For more leisurely signals, use the slowly varying circuit. The instrumentation amplifier can be constructed from three operational amplifiers. A filter circuit and a power supply complete the system.

168

Signal wires can introduce extraneous signals. They act like antennae, picking up electromagnetic energy, such as emanations from 60-cycle power lines, and then dumping it directly into your amplifier. To minimize the effect, keep the leads between the HET and the amplifier short. Additionally, you should use shielded wire. Lighton relies on three-core shielded cables. An electronics store may stock them, or you can make your own. Twist together three different colored wires, one each for the positive, negative, and signal leads of the HET. Wrap the wires inside a layer of aluminum foil and connect the foil to the circuit's ground with a short wire. For protection, add a layer of duct or electrical tape around the foil. The filter circuit provides another barrier to power-line noise. Finally, encase all your electronics inside a grounded metal project box.

You can read the output with a digital voltmeter or, better yet, use an analog-to-digital converter to record the data into a computer. For the most up-to-date suggestions, check out the Society for Amateur Scientists' web site at *www.sas.org*. Use shielded coaxial cable for the output connections, to prevent the HET from detecting the signal.

Lighton obtained some remarkable results after he superglued a whole rare-earth magnet to the abdomen of a *Blaberus discoidalis* nymph, a relative of the American cockroach. With the instrumentation amplifier's gain set to 100, the signal caused by the contractions of the dorsal vessel—the insect's heartbeats—is striking. After about 70 seconds of recording data, Lighton waved his hand in front of the nymph. The insect's heart stopped beating for several seconds. According to Lighton, that happened because the creature's nervous system may be too limited both to maintain circulation and to attend to stimuli.

Of course, any crawling by the insect will disrupt your results, so record data only when it is still. If the insect moves, it will generate a huge voltage signal that jumps well off the scale. In fact, Lighton reports that large signals occur whenever the insect opens its spiracles to breathe, about once every 5 to 30 minutes. By lowering the gain of your instrumentation amplifier, you can also monitor insect respiration.

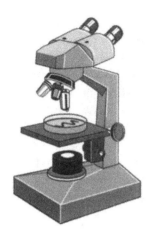

22 LEARNING IN SOW BUGS

by C. L. Stong, May 1967

Investigations of animal behavior have shown that dogs are easier to train than pigeons, and pigeons easier than frogs. Similarly, frogs exhibit more intelligence than fish. How far down the evolutionary ladder can one go before the line thins out between intelligent behavior, judged by an animal's capacity to learn from experience, and instinctive reaction?

Several experiments have been devised for investigating this question. One that amateurs can perform easily is designed to measure the learning response of invertebrates. The procedure is described by John Frost, a graduate student at California State College at Fullerton, as follows:

"An interesting specimen to use for observing learning behavior in invertebrates is the common sow bug, *Porcellio laevia*. These organisms live in moist places almost everywhere. The adult is about half an inch long. The body consists of seven free segments, each of which bears a pair of legs *[see Figure 22.1]*.

"The animal has no effective biological mechanism for preventing the evaporation of water from its body. In order to survive it must avoid the drying effect of direct sunlight. Hence it has learned to shun light. The experimenter can take advantage of this characteristic to train the bugs, causing each to run a simple maze in a direction contrary to the path preferred by the bug before training.

"Sow bugs can be found under rocks and logs. The insects may be scarce in winter and when the weather is hot and dry. In cities they tend to congregate during all seasons in the damp cellars of apartment buildings under wooden boxes and the like. If search is unsuccessful, try making a trap by hollowing out a potato and placing it under a tree or shrub. Cover

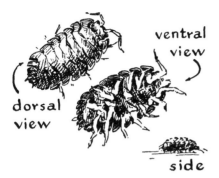

Figure 22.1 Views of a sow bug.

the potato with a few leaves and come back after 48 hours. The trap will usually contain several lively specimens. I do not recommend the potato trap for indoor use because it may attract some less desirable organisms. Sow bugs run when they are frightened, which is exactly what one wants them to do in the maze. They may faint when severely frightened, but they soon recover and scurry off.

"Captured sow bugs can be maintained indefinitely in a culture chamber improvised from a one-pound coffee can or a similar container. Half-fill the can with a mixture consisting of one part of sand by volume to two parts of leaf mold. On this surface place a peeled raw potato and a damp sponge of about the same size. Replace the potato and moisten the sponge every two or three days. The container should be closed by a perforated cover *[see Figure 22.2]*.

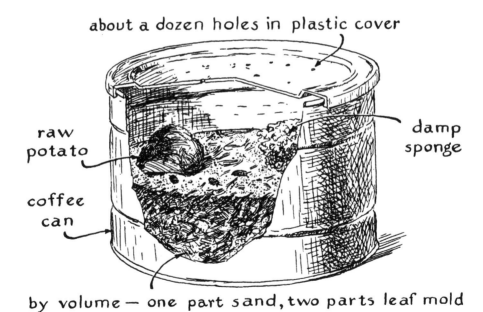

Figure 22.2 John Frost's container for maintaining the bugs.

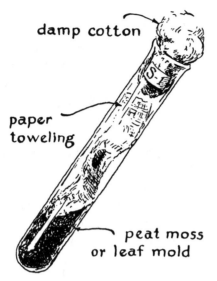

damp cotton

paper
toweling

peat moss
or leaf mold

Figure 22.3   Individual container.

"Specimens must be kept in individual chambers during training. These chambers can consist of test tubes half-filled with peat moss or leaf mold covered with a piece of paper toweling. A sliver of fresh potato is placed on the toweling along with the bug. The containers are loosely plugged with tufts of damp absorbent cotton and labeled so that the bug can be distinguished from others. Replace the potato as necessary and keep the cotton moist *[see Figure 22.3]*.

"The maze in which the bugs are trained consists of a simple *T*, made by cementing cardboard partitions in a box of clear plastic *[see Figure 22.4]*. The box should be about three inches wide, 4½ inches long, and ⅝ inch deep. The passages should be made about ⅜ inch wide. Rectangular openings that match the cross-sectional area of the

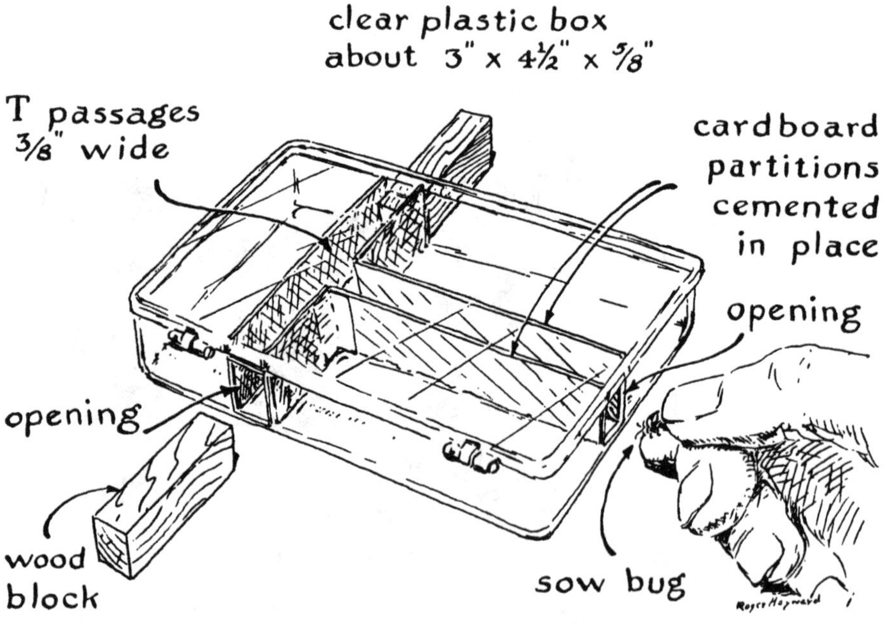

clear plastic box
about 3" x 4½" x ⅝"

T passages
⅜" wide

cardboard
partitions
cemented
in place

opening

opening

wood
block

sow bug

Figure 22.4   Apparatus for training bugs.

passages are cut in the walls of the box at the base of the *T* and at each end of the crossarm. Two blocks of wood that make a loose fit with the openings must be provided for closing either or both openings of the crossarm.

"The experiment is divided into two phases. First, determine and record the natural turning preference of each bug. Most sow bugs will take a preferred path through the maze. Having crawled up the leg of the *T*, some will habitually turn into the right portion of the crossarm and others into the left. Some will show no preference. During the second phase of the experiment the bugs are trained to turn in the direction contrary to their natural preference.

"Begin the experiment by transferring five or six specimens from the culture chamber to labeled individual chambers. Then remove a bug from a selected container and, holding it lightly between your thumb and forefinger, let it crawl from your fingertip into the base of the *T*. Record the direction of the turn, right or left. The bugs can tolerate only about 10 runs a day without suffering ill effects. For this reason the 20 runs needed to establish a reliable estimate of turning preference should span two days.

"Training is then accomplished by running each specimen through the course and punishing 'wrong' behavior. Each time a bug makes a turn in the direction it naturally prefers, immediately plug all exits with wood blocks and hold a 100-watt incandescent light close to the top of the passageway for about 20 seconds. When the bug turns in the direction opposite to its natural preference, plug the exit of the runway for 20 seconds but do not expose the animal to the punishing light.

"The training runs must be spaced a least five minutes apart. Between runs return the subjects to their individual quarters to 'think it over.' The experimenter can conserve time by training a number of animals sequentially. This practice also tends to increase the reliability of the experimental results and to minimize the statistical effect of the occasional specimen that does not survive the training experience.

"The training period should normally take three to 10 days, depending on how quickly the individual learns. The bugs should be subjected to no more than 10 training runs a day. At the conclusion of the training phase nine consecutive correct turns can be taken as evidence that the bug has learned. A correct turn is defined as one made in the direction opposite to the bug's natural preference as determined by the first phase of the experiment. Statistically it can be shown that nine consecutive correct turns will occur by chance only once in 100 runs.

"The procedure can be varied. For example, the omission of the bright light following a correct turn can be considered a reward. The leg of the *T* is lighted brightly until the bug reaches the crossarm; the light is removed if a correct turn is made. The desired behavior can be further reinforced by darkening the passage when the correct turn has been made.

"Much serious work has been done in recent years on the turning behavior of sow bugs as well as on the learning ability of cockroaches and box-elder bugs. The objective has been to clarify the role of reward and punishment in training procedures. I am certain that amateurs who repeat and extend the experiments will be surprised to find evidence of intelligent behavior so low on the scale of evolution. They will also have the satisfaction of exploring animal behavior by means of experimental procedures that do not injure the organism."

PART 5

MICROBIOLOGY

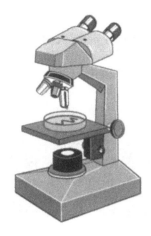

23 MOUSE GENETICS

by Albert G. Ingalls, December 1952

Four years ago, when Marita Mullan of Philadelphia, Pennsylvania, was 15 years old, a boy who knew she liked animals gave her a pair of common house mice. The joke backfired. Her growing enthusiasm for the new pets soon left little time for him. She started haunting the public library and registered for special courses at the University of Pennsylvania, though she was too young to receive credit. Within a little more than two years she was invited to address the scientific staff of the Jackson Memorial Laboratory in Bar Harbor, Maine. With the help of her mice, Miss Mullan had become an advanced amateur geneticist.

Miss Mullan's interest in genetics was sparked by a chance observation. One morning, when she went to the basement to water her mice, she made a strange and, as she learned later, rare discovery. A few days earlier the mice had produced a litter of six young, all pink and hairless. Now they were growing coats. This morning she noticed that one of the baby mice was not a mousy gray, like its parents, but bright orange! In the course of searching for the answer she bred mice of many colors and learned why the average householder is unlikely ever to trap an orange one.

She began with a review of the work of genetics' most celebrated amateur: the Austrian monk Gregor Johann Mendel. Mendel also started with a question. What factor, he wondered, relates true-breeding varieties within a species? For the answer he put the question directly to nature. He had had no previous training in science and so was forced to discover its method for himself: he stated his problem, experimented, observed, theorized, validated his theory by further experiment, and finally expressed his findings as a set of natural laws.

Although Mendel worked with peas his basic laws of heredity hold equally for mice, oak trees, or *Homo sapiens*—for any organism that reproduces sexually. He cultivated and crossbred two varieties of peas—tall and dwarf. When the offspring of these matured, he was surprised to observe they had all grown tall, like one of the parents, instead of medium-sized as he had expected. He then interbred these tall hybrids. Again he was surprised. This time the offspring were both dwarf and tall—and in the precise ratio of three tall plants to one dwarf.

What set of circumstances, Mendel wondered, would produce the orderly result he had observed? After much pondering he finally hit upon a theory that goes like this: Suppose the cells of the parent plants each contain a pair of factors in the form, for instance, of minute particles. One kind of particle can cause a plant to grow tall and the other dwarf. Suppose further that when the male germ cell of one plant unites with the female germ cell of the other, each contributes only one member of its pair of particles to the seed. The factor from this male parent, let us say, is invariably of the type causing tallness, and from the female, dwarfism. Then all members of the first hybrid generation from this pair of parents would get a tallness factor from the "father" and a dwarfism from the "mother." Assume that the tallness factor is dominant. Then all the offspring would grow tall.

But when these hybrids interbred in turn, a different result would be expected. The inheritance could now be mixed in four different combinations: tallness plus tallness, tallness plus dwarfism, dwarfism plus tallness, dwarfism plus dwarfism. The first three of these combinations would produce a tall plant, the last a dwarf plant. So on the average three out of four of the second generation should be tall.

It was a clever explanation—but did it really describe nature? Mendel tested it statistically on various hereditary characteristics of peas—the shape and color of the pods, the position of the flowers, the length of the stems, and the texture of the seeds—and found that in every experiment his theory accurately predicted the results. Today Mendel's mysterious factors are called genes.

Every organism begins life as a single cell. The genes are found in the cell's nucleus, arranged in threads or filaments like strings of beads. The threads are called chromosomes, from the fact that they can be stained with colored dyes for observation under the microscope. When a cell prepares to divide, the barely visible chromosomes gradually thicken and finally split down their length, each of the many genes in each new piece being duplicated in the other. Thus each of the daughter cells inherits a duplicate set of chromosomes and genes *[see Figure 23.1]*.

In all of nature the gene is the only known structure with the power to manufacture an exact copy of itself. But this does not mean that a gene can-

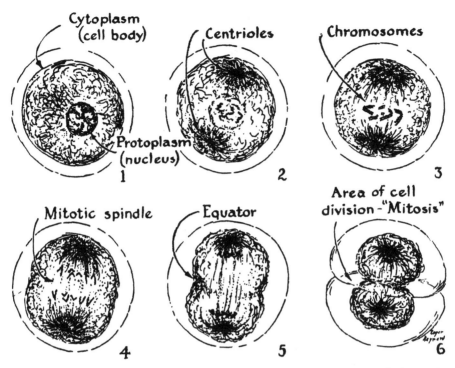

Figure 23.1 Cell division begins when chromosomes divide and are pulled apart by filaments extending from the centrioles. The process ends when the cell splits into two daughter cells, which then mature and repeat the whole cycle.

not be changed. Every now and again some force acts on one gene or another and alters it. X rays will do this. So will cosmic rays, various chemicals, extremes of temperature, and other influences. Unless the affected gene kills its host cell, it will subsequently go right on making copies of itself—in its new, altered form. Such modified genes are called mutants. They account not only for the varieties within the species but also provide the basis for the origin of new species through natural selection. Some mutations are beneficial to the organism's chances of survival, but most are harmful. Fortunately most of the desirable ones are dominant and hence assist the development of useful traits. In contrast, recessive genes carrying a potential of undesirable characteristics can express themselves only when chance happens to pair them with like recessives in an offspring. In a large population the chances of such a meeting are small. Hence many generations may come and go before the unfortunate trait appears.

Marita Mullan's orange mouse represented such a rare event. The gene responsible for the trait is known as a "lethal yellow." As the name implies, this gene carries other changes more serious than the orange color. Lethal yellow mice die before they are old enough to mate. Miss Mullan's orange mouse died within a few weeks.

Such lethal mutants appear in all species. Fox breeders, for example, cultivate a highly prized variety of "platinum" fox. When platinum foxes mate, the hybrid is a snowy white. Like orange mice they never survive. Several lethal genes are carried by man, and science has learned how to circumvent the effects of some. Diabetes is genetic in origin; it is due to a defect or absence of the gene responsible for the manufacture of insulin. This genetic failure must be offset by administering insulin artificially.

The treatment of diabetes provides a clue to the nature of the chemical mechanism through which genes express themselves. The genes apparently govern the complex chemistry of cells, each gene being responsible for the cell's ability to manufacture a particular chemical link, called proteins. This means that a great number of different kinds of chemicals must bathe the center of the cell, because complex organisms such as mice and humans exhibit thousands of different traits, each accounted for by its own unique gene and corresponding protein. It follows also that in this conglomerate chemical stew all the genes must to some extent modify one another's effects. Few genes, if any, act in complete independence.

In view of the interaction of the genes, it is logical to suppose that their position on the chromosomes could play a major role in shaping inherited traits. This is indeed the case, and it accounts for another mechanism by means of which varieties arise within a species. Sometimes when the new cell makes its initial division, the chromosomes break and grow together again in a different order, or several may break and exchange sections during reassembly. Short lengths, together with their complement of genes, may even be lost in the process. Any of these and related chance happenings may result in an offspring which carries a genetic structure differing from that of the parents. Hence, the offspring may possess a set of characteristics strange to the lineage—some visible and others obscure.

After a time, particularly in large populations where individuals mate outside the family line, the most dominant traits emerge. All individuals within the species bear marked superficial resemblance to one another, although the genetic systems of the mating partners may carry hundreds, even thousands, of contrasting and hidden recessives. Many of the recessives may be endowed with potential control over some one trait, such as hair color, but may never express themselves until they become paired with others of like kind through chance mating. Thus variation in many traits is subject to the control of a whole group of genes.

It was with such a group that Marita Mullan worked. She writes:

"Little did I realize the storehouse of potential beauty that my original pair of mice were hiding in the form of recessive mutants. Gradually, however, many of these varieties appeared in the offspring. Most of them are well known. Some were described long before the time of Christ by the Chinese, who bred these animals because of their singular beauty. Several unlisted traits came to light during my experiments.

"The most treasured of all fancy genes, and the most beautiful, is the pink-eye dilution. This gene somewhat reduces the amount of pigment in the skin and eyes. It also tends to produce a smaller mouse. The maltese or blue dilution is another mutant that tends to reduce the amount of pigment, but less drastically than the pink-eye gene. Black and brown is another striking combination and produces a blue-gray coat and a very beautiful chocolate color.

"The basic colors of the mouse's fur are produced by what is called the agouti series. 'Agouti' means that the fur has a characteristic variegated appearance, caused by the fact that each hair has a light and dark portion. It accounts for the typical mousy appearance of the wild mouse's fur. The series also contains the light-bellied agouti, the black-and-tan, and the so-called nonagouti or black."

The work of a geneticist differs from that of a person who simply breeds animals or plants. In the first place, the geneticist hopes to learn more about the mechanism of heredity and, if possible, about the scientific basis of life itself. In the second place, the geneticist's breeding technique is guided by tested and proved laws, while the breeder generally proceeds on the rule of thumb that "like produces like."

Breeding by this classical method has gradually improved the stock of many plants and animals. By selective mating of individuals exhibiting the desired traits, people have developed products ranging from rust-resistant wheat to race horses. But after a time this method reaches a limit: no amount of careful selection seems to make any additional improvement. The quality of the stock levels off.

Genetics can do much better. At the turn of the twentieth century midwestern farmers were content with a strain of corn, for example, which yielded about 25 bushels per acre on the average. Today hybrid corn, developed by scientific application of principles of genetics, produces yields exceeding 200 bushels per acre!

How does genetics achieve such sensational improvement? Primarily by close inbreeding to establish desired traits and then the crossing of two unrelated inbred strains. Generally the product of this cross shows amazing qualities, far surpassing those of the immediate parents and of the ancestors on each side. Geneticists call the effect "hybrid vigor."

These and related principles guided the work of Marita Mullan. From her original pair of wild mice she developed dozens of independent strains.

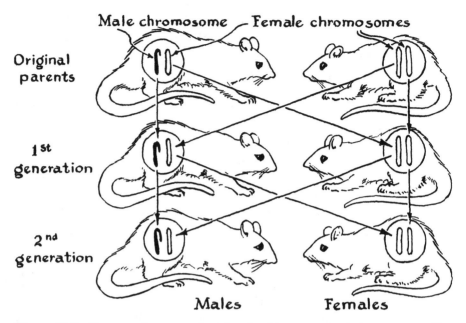

Figure 23.2 The sex of a mouse is determined by one pair of chromosomes. The hooked black chromosomes in this diagram bear dominant factors for maleness. The straight white chromosomes bear recessive factors for femaleness.

She kept careful records of each individual and, following the example of Mendel, made tables showing the number of individuals in each generation with like characteristics. Her tables were far more complex than those listing the two factors of tallness and dwarfism in peas, however, because the several colors in mice are determined not by two genes but by a series. This vastly increases the number of possible combinations and adds to the interest and challenge of the game *[see Figure 23.2]*.

As Miss Mullan says, "The thrill of breeding unusual offspring is not the only appeal of genetics. Those who have crossword-puzzle minds will find that genetics on paper becomes a fascinating and challenging form of mental gymnastics. A simple knowledge of the kinds of genes and how they are distributed on chromosomes is all that one needs to commence dreaming up problems of inheritance and writing down the specifications for the new kind of individual you wish to withdraw from nature's reservoir. The chance combinations in this reservoir are not limited to color in mice. The study of the structural abnormalities of the skin and fur, for example, can be exciting—and sometimes amusing.

"The most comical of all these mutants is the hairless. The hairless mouse spends the first two weeks of its life growing a full normal coat of fur and at this point cannot be distinguished from its normal brothers and sisters. Soon, however, the fur begins to drop out and the hair line swiftly recedes to complete baldness, so that in a few days the young mouse resembles a bald man in a fur coat. Shortly the top of this coat is shed and the mouse seems to be wearing breeches. The final stage is perhaps the most amusing of all, for in a few days all trace of hair is gone except for a fringe about the haunches, and the mouse looks for all the world like a small, awkward ballerina. The entire loss is comparatively rapid and is completed in about 14 days. Thus four weeks after birth the creature has grown a coat and lost it—is finally as naked as a newborn baby. If there are no complicating factors, the mouse will soon regenerate a coarse fuzz which usually remains throughout its life. The first mice of this kind were caught in London in 1926.

"Some mice are not totally hairless, and yet are not completely furred. These have long, fine fur which is much less dense than that of the normal mouse. In some the length of the fur is so reduced that it is necessary to use a magnifying glass to examine the quality of the fur.

"These strange characteristics are but a few of the interesting mutants which have appeared as the result of breeding two apparently uninteresting mice."

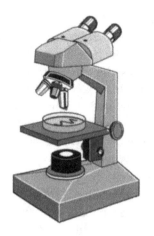

24 FRUIT FLY GENETICS

by C. L. Stong, June 1965

In 1951 Ernst Hadorn, who is now rector of the University of Zurich, and Herschel K. Mitchell of the California Institute of Technology conducted an interesting experiment involving the genes responsible for eye color in the fruit fly *Drosophila*. The experiment showed that the genes also control the production of a family of compounds that in effect serve to label the flies by genetic type. The compounds were isolated by the analytical procedure of paper chromatography. Hadorn and Mitchell crushed anesthetized flies on a piece of filter paper, one edge of which was subsequently immersed in a solvent. As the solvent migrated through the fibers of the paper it washed the compounds from the spot made by crushing the flies. When the paper was viewed under an ultraviolet lamp, the compounds appeared as a series of fluorescing bands spaced according to the solubility of the compounds and to differences in their affinity for the paper. Subsequently the compounds were identified as pteridines, a name that comes from "pteron," the Greek word for "wing"; the first compounds of the family were found in the wings of butterflies.

A modified version of the experiment that can be done at home has now been worked out by Richard LaFond of Monson, Massachusetts. Although LaFond's apparatus is largely assembled from scrap materials and presents a deceptively simple appearance, it provides the experimenter with a powerful means for delving into an exciting aspect of genetics. LaFond writes:

"The eye of *Drosophila* has been found to contain two genetically controlled pigment systems, one brown and the other red. These systems were first revealed by their different solubilities. The red pigment is water-soluble but the brown is not. During the fly's early development the brown pigment appears first; the red, some hours later. The normal eye color of flies of the

184

"wild" type, such as Oregon-R, is brick red, caused by the presence of both pigments. This eye color appears when all genes are working normally.

"In the case of mutant flies that have eyes of abnormal color, such as scarlet, a gene suppresses the formation of brown pigment. Accordingly the eyes are red. Mutants with brown eyes, on the other hand, have a gene that suppresses the formation of red pigment. A cross between a mutant with scarlet eyes and one with brown eyes produces a hybrid with white eyes. In effect the pigment systems cancel out.

"It is the red pigments and other brightly fluorescing compounds that comprise the pteridines. These compounds are situated not only in the eyes of the fly but also in the ovaries and testes and in the Malpighian tubules, which act as a kidney. The relative amounts present in a specimen tend to differ at each stage of the life cycle as well as between mutants and their hybrid offspring. For this reason experiments having to do with the pteridines are open to almost limitless variation.

"One must, of course, have a stock of flies in order to conduct experiments. An easy way to collect *Drosophila* is to leave outdoors in a shaded area a culture bottle containing a special food rich in yeast. By careful inbreeding it is possible to develop a number of mutant strains from the wild stock. Specimens of all types also can be bought from suppliers. My initial flies were obtained from the Curator of Stocks, Division of Chemotherapy, The Institute for Cancer Research, 7701 Burholme Avenue, Philadelphia, PA 19111. [You can also obtain flies from WARD'S Natural Science Establishment; 800-962-2660 or *www.wardsci.com*. Ed.]

"Having acquired a small initial stock by capture or purchase, the experimenter then perpetuates the stock by culturing techniques. The live specimens come in small vials. Adults are promptly transferred to a culture bottle, but the vials are not discarded immediately. Eggs have been laid in the food from which young flies will soon hatch.

"I use half-pint milk bottles as culture vessels. Before transferring flies to these containers each bottle is sterilized and equipped with a supply of food. A number of food preparations have been developed for culturing *Drosophila*. I use the recipe devised by Boris Spassky of the Rockefeller Institute. This nutrient is made by adding 194 milliliters of tap water to 29 milliliters of unsulfured molasses and bringing the mixture to a boil in a pan. To the boiling solution are added 26 grams of regular Cream of Wheat and 2 grams of uniodized salt. The mixture must be stirred constantly and cooked for about five minutes. The pan is then removed from the stove. Two milliliters of a 10 percent solution of Tegosept M, a brand of methyl-p-hydroxybenzoate [also available from WARD'S], are stirred into the mixture as a preservative. The 10 percent solution is made by diluting 10 grams of the compound in 100 milliliters of 95 percent ethyl alcohol.

"To milk bottles that have been thoroughly washed and boiled in water add the food mixture to a depth of about half an inch by means of a funnel that prevents food from spattering on the glass. Wipe any condensed water from the inner wall of the bottle. Plug the opening of each bottle with an unwaxed paper cap or a tuft of cotton covered with a piece of cheesecloth. Place the bottles in a pressure cooker containing about 100 milliliters of water and boil for 30 minutes at a pressure of 15 pounds per square inch and a temperature of 120 degrees C.

"After sterilization stand the capped bottles on a convenient shelf until any large drops of water adhering to the inner wall evaporate. This step is important because in an excessively moist bottle the flies may get stuck in the food medium and drown. Excessive moisture can be removed with a sterilized paper towel. When the interior of the bottle is dry, fold a piece of sterilized paper toweling 11 inches long and 2½ inches wide into quarters, so that the folded sheet measures 2¾ by 2½ inches, and push one end below the surface of the food medium. The paper strip provides a place on which *Drosophila* larvae can pupate. *Drosophila* thrive on a fermenting medium. This is provided by sprinkling a pinch of active dry yeast over the surface of the food. Each bottle is then carefully labeled with the name of the type of fly it will house and the date *[see Figure 24.1]*.

"Cultures of *Drosophila* should not be kept more than 20 days because they may become infected with mites, which markedly decreases their abundance. Mites are minute members of the class Arachnida; an effective

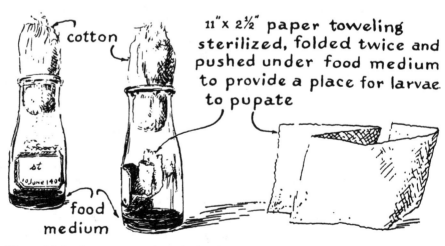

cotton

11" x 2½" paper toweling sterilized, folded twice and pushed under food medium to provide a place for larvae to pupate

food medium

Figure 24.1 Arrangements for culturing fruit flies.

agent against them is benzyl benzoate. [This chemical can be purchased from the Society for Amateur Scientists. Ed.] Make up a solution of this compound and mineral oil in equal proportions. Shake the solution well and spread it on the shelves supporting the culture bottles. Maintain a temperature of 72 degrees Fahrenheit in the storage room. A new generation of flies will appear in 12 to 14 days. When discarding an old culture, always wash the used bottles thoroughly in hot water and any convenient detergent and then boil them in fresh water to destroy mold spores.

"Before you count or handle individual flies you must anesthetize them. The anesthetizing apparatus consists of a peanut butter jar closed with a large cork through which the spout of a small funnel is inserted. A small sponge in the bottom of the jar is moistened with a few drops of ether. A miniature cage for holding the flies is made from the larger section of a No. 000 gelatin capsule (sold by Torpac, 973-244-1125, *www.torpac. com*) by perforating the gelatin about six times with a hot sewing needle. The open end of the capsule is slipped partway into or over the end of the funnel and, if it does not make a snug fit, taped in place. Flies are transferred to the anesthetizing cage by placing the mouth of the open culture bottle tightly against that of the funnel, orienting the assembly so that the bottle containing the ether is on the bottom and tapping the culture bottle. The flies will fall down through the funnel into the anesthetizing cage. I use Merck motor ether. The container in which the ether is stored must be tightly closed when not in use *[see Figure 24.2]*.

"The flies must be counted as one step in nearly all experiments. It is also often necessary to separate them by sex, type, age, and so on. During

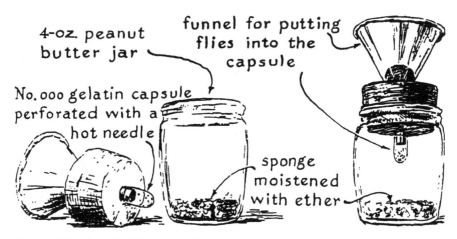

Figure 24.2 Anesthetizing apparatus.

such steps the anesthetized specimens are manipulated by means of a small camel's-hair brush, preferably on a smooth white surface such as white glass or a sheet of clear glass that rests on white paper. The sex of specimens is easily determined by examining inverted flies under a five-power magnifying glass. The distinguishing sexual features appear in the accompanying illustration *[see Figure 24.3]*.

"Normally flies remain anesthetized for 5 to 10 minutes. Some individuals revive sooner than others. These can be anesthetized again by inverting over them a petri dish or other shallow container into which is fastened a small piece of paper toweling moistened with a few drops of ether. Remove specimens from the ether promptly when they stop moving. Overexposure will kill them.

"After a culture has been maintained for 20 days transfer all adults to a fresh culture bottle. Recap the old bottle and 48 hours later transfer the young flies that have hatched during the interval to a fresh culture bottle. The old flies can be used to start new cultures. Develop scrupulously clean work habits in order to avoid contaminating or mixing cultures. Specimen types can be mixed accidentally, for example, by transferring a soiled glass rod or other implement to which an egg adheres from one bottle to another.

"For separating the pteridines I use a chromatographic apparatus of the descending-paper type. Essentially it consists of a closed glass box that houses an elevated container of solvent in which the upper end of the paper is immersed. The dimensions of the apparatus are shown in the accompanying illustration *[see Figure 24.4]*. A square inch of glass is cut from one corner of the close-fitting cover to provide access for transferring the solvent to the container. This opening is sealed with a thick sheet of paraffin in which a round hole about half an inch in diameter is made. The

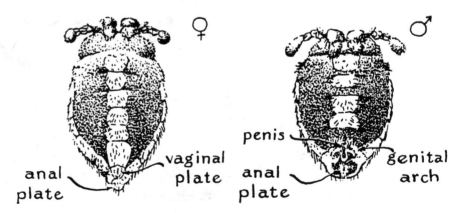

Figure 24.3   Ventral view of fruit flies: female (left) and male (right).

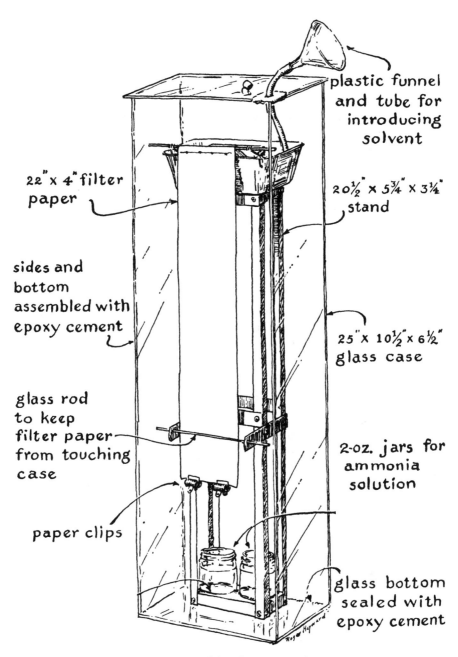

plastic funnel and tube for introducing solvent

22" x 4" filter paper

20½" x 5¾" x 3¼" stand

sides and bottom assembled with epoxy cement

25" x 10½" x 6½" glass case

glass rod to keep filter paper from touching case

2-oz. jars for ammonia solution

paper clips

glass bottom sealed with epoxy cement

Figure 24.4 Dimensions and parts of the chromatograph.

hole is then fitted with a removable stopper, which is also made of paraffin. The container for the solvent, which rests on the top platform of a removable framework, can be any convenient shallow vessel about seven inches long and an inch or two deep. I use an aluminum pan that rests on a framework made of parts from an old Erector set.

"Brackets of wire and paraffin, attached to one edge of the pan, support a slender glass rod 10 inches long over which a piece of moist filter paper is draped. In addition to serving as a support for the paper, the rod prevents a siphoning action that would cause the solvent to flow; the only kind of flow should be that due to capillary action. The upper edge of the paper strip is weighted against the bottom of the solvent container by a glass butter dish filled with sand held in place by a layer of melted paraffin. A second glass rod, attached about halfway down the framework, serves as a stop to keep the paper away from the glass housing.

"Two glass jars of about 50-milliliter capacity are placed on the lower platform of the framework. These each contain 30 milliliters of a solution that by evaporation brings about an equilibrium between the atmosphere of the chamber and the vapor content of the filter paper. One can make a 100-milliliter stock of this solution by diluting 25.9 milliliters of 27 percent ammonium hydroxide with distilled water. The entire apparatus must be carefully leveled before use so that the surface of the solvent in the container makes a right angle with respect to the centerline of the paper strip *[see Figure 24.5]*.

"Chromatograms are made on strips of Whatman No. 1 chromatography paper cut 4 inches wide and 22 inches long. My paper was bought from Howe and French, 99 Broad Street, Boston, Mass. 02110. [It's also available from WARD'S Natural Science Establishment; 800-962-2660 or *www.wardsci.com*. Ed.] It comes in sheets 18¼ inches long by 22½ inches wide. As an aid in placing specimens uniformly on the paper I draw a light pencil line squarely across each strip at a distance of 6¾ inches from one end and divide the line into five equal intervals by four light pencil dots. The material to be analyzed is placed on these dots.

"To prepare for the analysis of adult flies first anesthetize selected specimens of the same age. Age is an important factor because the concentration of pteridines in the flies varies during the life cycle. The concentration also differs substantially between the head and the body. Moreover, the chromatograms of males differ from those of females.

"One begins a typical experiment by severing the heads of 10 flies with a razor blade and squashing the material onto the paper with the end of a glass rod. I always reserve the right-hand dot for the control specimen, which is prepared by applying to the dot the heads of 10 Oregon-R wild-

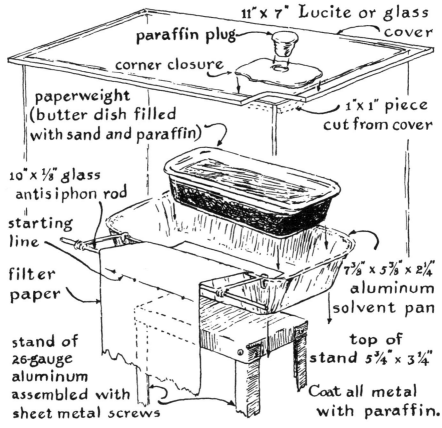

Figure 24.5 Details of the chromatograph.

type flies. If the control specimen fails to separate as anticipated, the chromatogram is discarded. The control also provides a convenient cross-check for estimating the amounts of pteridines in other specimens in relation to those naturally present in the wild type.

"When one is making chromatograms of fly bodies rather than heads, one must take care to separate all head tissue cleanly. A small amount of head tissue can contain more pteridines than a whole body, hence even a tiny fragment can seriously distort a body chromatogram. The bodies are boiled in water for three minutes to coagulate the protein and thus facilitate the chromatographic separation. After boiling they are placed on paper toweling to remove excess water. Five bodies are applied to each spot. All spots must be dried at room temperature before the chromatographic paper is placed in the apparatus. In addition, a pencil notation is made next to each spot that includes its descriptive initials—such as *p* for plum eye, *st* for scarlet or bright red eye, *v* for vermilion eye, *w* for white-eye mutants, *bw* for brownish eye, *w^a* for white-apricot eye—together with the date and time.

"After the paper strip has dried, the solvent pan is placed on the upper platform of the framework. The strip is then draped over the upper glass rod and anchored in place by the butter dish. The loose end of the strip is threaded between the lower glass rod and the framework so that it hangs freely suspended. Paper clips of the pinch type are attached to the bottom edge as weights. The ammonium hydroxide containers are then put on the lower platform. Now the entire assembly is placed in the glass housing, covered by the glass top, draped with a cloth that excludes light, and left undisturbed for two hours. During this interval the vapor content of the paper reaches equilibrium with the atmosphere of ammonia released by the ammonium hydroxide. Good separations cannot be expected if the interval of equilibration is stinted.

"When equilibration is complete, remove the cloth cover and the paraffin plug and insert into the solvent container a rubber tube equipped with a small plastic funnel. This is used to transfer 230 milliliters of developing solvent into the tray. The solvent consists of 360 milliliters of 1-propanol (n-propyl alcohol), 45 milliliters of 7 percent aqueous ammonia, and 135 milliliters of distilled water. Finally, remove the rubber tubing, reinsert the paraffin plug promptly, cover the apparatus to exclude light, and maintain the room temperature at 68 degrees F. for 20 hours. At the end of this interval remove the chromatogram from the apparatus. Handle the paper by its ends and suspend it upside down for drying at room temperature. Return the solvent and equilibrating solutions to their respective storage bottles.

"When the chromatogram has dried, it can be examined under an ultraviolet lamp. If all has gone well, the characteristic fluorescent bands will appear *[see Figure 24.6]*. Caution: Never look directly at the bulb of an ultraviolet lamp. The rays are injurious to the eyes. Protective goggles should always be worn when working with ultraviolet radiation. Dual lamps that emit ultraviolet at wavelengths of 2,537 and 3,660 angstrom units and operate from regular house current are available from the Edmund Scientific Co. of Tonawanda, NJ, 800-728-6999, *www.edsci.com*. Pteridines that emit blue or violet light glow with greatest brilliance when they are irradiated at a wavelength of 3,660 angstroms; those that emit reds or yellows appear brightest when they are irradiated at a wavelength of 2,537 angstroms.

"Chromatograms should be examined and evaluated as soon as they dry because the fluorescent compounds tend to fade with time. As a convenience in scoring and recording the results I prepare a table in advance. Symbols designating all specimens are listed in a column that extends

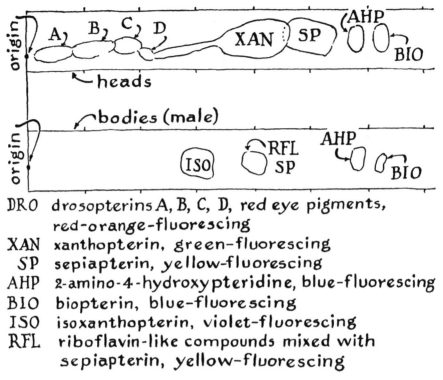

DRO   drosopterins A, B, C, D, red eye pigments,
      red-orange-fluorescing
XAN   xanthopterin, green-fluorescing
 SP   sepiapterin, yellow-fluorescing
AHP   2-amino-4-hydroxypteridine, blue-fluorescing
BIO   biopterin, blue-fluorescing
ISO   isoxanthopterin, violet-fluorescing
RFL   riboflavin-like compounds mixed with
      sepiapterin, yellow-fluorescing

Figure 24.6   Diagram of a chromatogram.

down the left edge of the page *[see Figure 24.7 on page 194]*. A similar row across the top of the page lists the several pteridines: DRO for drosopterins that fluoresce orange-red; XAN for xanthopterin, green; SP for sepiapterin, yellow; AHP for 2-amino-4-hydroxypteridine, blue; BIO for biopterin, blue; ISO for isoxanthopterin, violet, and RFL for riboflavinlike compounds mixed with sepiapterin, yellow. The apparent brilliance of each band as judged by eye is then recorded by means of plus and minus symbols, as shown in the accompanying illustration.

"Isoxanthopterin is present in the bodies of both male and female flies but is found in much larger amounts in male bodies because it is concentrated in the testes. This difference in concentration, as disclosed by the chromatogram, immediately identifies the sex of most specimens. Certain mutants, however, such as the rosy-1 (ry^1), rosy-2 (ry^2) and maroonlike (*ma-l*), do not produce detectable quantities of isoxanthopterin. Rosy-2 males can still be distinguished by an abnormally large amount of 2-amino-4-hydroxypteridine, the precursor compound in the formation of isoxanthopterin.

Mutant	Drosopterins				XAN	SP	AHP	BIO	ISO	RFL
	A	B	C	D						
p	++	++	++	++	+	++	+	++	+++	++
st	+++	+++	+++	+++	+++	+++	++	++	+++	++
v	+++	+++	+++	+++	++	++	++	++	+++	++
Ore.-R	+++	+++	+++	+++	+++	++	++	++	+++	++
w	−	−	−	−	−	−	−	−	−	−
bw	−	−	−	−	−	−	−	−	−	−
w^a	−	+	+	+	+	+	+	+	++	+
ry	++	++	++	++	++	+++	++++	+++	−	++
wild*	+++++	+++++	+++++	+++++	+++	++	+++	+++	++	++
se x w**	+++	+++	+++	+++	+++	+++	+++	+++	+++	++

* D. willistoni

** Sepia female
crossed with
white male

− = nothing apparent
+ = trace
++ = small amount
+++ = moderate amount
++++ = large amount
+++++ = very large amount

Figure 24.7 A method for charting the results of chromatograms.

"Interesting changes in the amounts of the pteridines with time can be observed by doing chromatography of larvae, young and old pupae, and hatched flies of various ages. Young pupae are brown and translucent, whereas older pupae are opaque. Larvae that have begun to climb onto the paper in the culture bottle and up the wall of the bottle are removed with a glass rod and applied directly to the paper. I chromatographed five larvae, five pupae, and ten heads of sepia mutants one day old. The larvae show a very weak fluorescence that does not seem to indicate any particular pteridine. Young pupae display xanthopterin, together with a substance I could not identify that fluoresces in the blue portion of the spectrum. There was also a hint of yellow sepiapterin. Xanthopterin and sepiapterin appear in substantial amounts during the late pupal stage.

"The experiments I found most interesting involved crossing different eye color mutants (and also crossing such mutants with flies of the wild type) and then chromatographing the offspring. Virgin females must be used for making controlled crosses because females can store the sperm from one insemination for a large part of their reproductive lives. To collect virgin females, clear a culture bottle of all flies. Search the paper care-fully for adults that may be hidden in its folds. Use a bottle that contains many pupae from which new flies will soon hatch. From this bottle collect females within 10 hours of the time they hatch. (After 10 hours they will mate with males, although they do not lay eggs for two days.) Transfer the virgins to a fresh culture bottle placed on its side so that they will not stick to the food at the bottom. Males of any age and desired type are then anes-thetized for transfer. Several pairs of males and females are placed in a fresh culture bottle and labeled with the description of the cross and the date. When larvae begin to climb up the wall of the bottle and onto the paper, the adults are removed.

"Always remember when chromatographing young flies of any type that the relative amounts of the pteridines change with the age of the spec-imen. Use specimens that are approximately the same age. I usually select flies that are 12 days old. To age specimens, clear a culture bottle of adults, collect the young as they hatch during an interval of an hour or so, and then transfer the young to a fresh, dated culture bottle. Chromatograph after 12 days.

"An important consideration in interpreting the chromatograms of the mutants and the crosses is whether the characteristic pattern is the result of the major gene or genes under consideration or merely a reflection of the specimen's genetic background. Genes other than those assumed by the experimenter can modify the amounts of pteridines in the fly. For example, it has been shown that the amount of isoxanthopterin is influ-enced by many genes because there is less variation in the concentration of this compound in inbred stocks of flies than in flies that have been col-lected in nature and mass-cultured for maximum heterozygosity, or varia-tion of genes.

"In my experiments I have assumed that the obvious differences between chromatograms of the various mutants are associated with a major gene. Although I chromatographed many flies, I made no attempt to exclude the possibility of background effects by crossing mutants with flies of the wild type and then reisolating the stock after six or seven backcross generations to compare the chromatographic results with those of the original strain. This step would be essential, however, in a quantitative study designed to reveal the influences of genetic background."

25 SPOOLING THE STUFF OF LIFE

by Shawn Carlson, September 1998

I f you're a man who's looking to get married, here's some friendly advice. Only consider women who are smarter than you are. I followed this prescription in 1994 when I married Michelle Tetreault, a charming and brilliant biophysicist. Now my wife always intrigues me with her insights and never lets me get away with anything dubious at home.

At work Michelle employs the latest techniques in biochemistry to unravel the secrets of photosynthesis. By manipulating the smallest units of inheritance, the individual base pairs on a single strand of DNA, she can change one by one the amino acids that make up a key protein and then study how well this altered molecule can do its job.

Hearing about Michelle's research so often at the dinner table recently prompted me to try my hand at molecular biology. Although the many cutting-edge techniques she uses are probably beyond the range of amateur dabbling, recent advances have opened up intriguing avenues for informal explorations into biotechnology. To help clear the way, this chapter explains how anyone can do what biotechnologists do routinely: extract and purify DNA.

DNA is the largest molecule known. A single, unbroken strand of it can contain many millions of atoms. When released from a cell, DNA typically breaks up into countless fragments. In solution, these strands have a slight negative electric charge, a fact that makes for some fascinating chemistry. For example, salt ions are attracted to the negative charges on DNA, effectively neutralizing them, and this phenomenon prevents the many separate fragments of DNA from adhering to one another. So by controlling the salt concentration, biologists can make DNA fragments either disperse or glom together. And therein lies the secret of separating DNA from cells.

The procedure is first to break open the cells and let their molecular guts spill into a buffer, a solution in which DNA will dissolve. At this point, the buffer contains DNA plus an assortment of cellular debris: RNA, proteins, carbohydrates, and a few other bits and pieces. By binding up the proteins with detergent and reducing the salt concentration, one can separate the DNA, thus obtaining a nearly pristine sample of the molecules of inheritance.

My profound thanks go to Jack Chirikjian and Karen Graf of Edvotek, an educational biotech company in West Bethesda, Maryland, for showing me how anyone can purify DNA from plant cells right in the kitchen. You'll first need to prepare a buffer. Pour 120 milliliters (about 4 ounces) of water into a clean glass container along with 1.5 grams (¼ teaspoon) of table salt, 5 grams (one teaspoon) of baking soda and 5 milliliters (one teaspoon) of shampoo or liquid laundry detergent. These cleaners work well because they have fewer additives than hand soaps—although do feel free to try other products as well.

The detergent actually does double duty. It breaks down cell walls and helps to fracture large proteins so they don't come out with the DNA. The people at Edvotek recommend using pure table salt and distilled water, but I have used iodized salt and bottled water successfully, and once I even forgot to add the baking soda and still got good results. In any case, try to avoid using tap water. To slow the rate at which the DNA degrades, it's best to chill the buffer in a bath of crushed ice and water before proceeding.

For a source of DNA, try the pantry. I got great results with an onion, and the folks at Edvotek also recommend garlic, bananas, and tomatoes. But it's your experiment: choose your favorite fruit, vegetable, or legume. Dice it and put the material into a blender, then add a little water and mix things well by pulsing the blades in 10-second bursts. Or, even simpler, just pass the pieces through a garlic press. These treatments will break apart some of the cells right away and expose many cell walls to attack by the detergent.

Place 5 milliliters of the minced vegetable mush into a clean container and mix in 10 milliliters of your chilled buffer. Stir vigorously for at least two minutes. Next you'll want to separate the visible plant matter from the molecule-laden soup. Use a centrifuge if possible. Spin the contents at low speed for five minutes and then delicately pour off at least 5 milliliters of the excess liquid into a narrow vessel, such as a clean shot glass, clear plastic vial, or test tube. If you do not have a centrifuge, strain the material through an ordinary coffee filter to remove most of the plant refuse. With luck, any stuff that leaks through should either sink or float on top, so it will be a simple matter to pour off any solids into the sink and then decant the clear liquid into a clean vessel.

The solution now contains DNA fragments as well as a host of other molecular gunk. To extract the DNA, you will need to chill some isopropyl alcohol in your freezer until it is ice-cold. Most drugstores sell concentrations between 70 and 99 percent. Get the highest concentration (without colorings or fragrances) you can find. Using a drinking straw, carefully deposit 10 milliliters of the chilled alcohol on top of the DNA solution. To avoid getting alcohol in your mouth, just dip the straw into the bottle of alcohol and pinch off the top. Allow the alcohol to stream slowly down along the inside of the vessel by tilting it slightly. The alcohol, being less dense than the buffer, will float on top. Gently insert a narrow rod through the layer of alcohol. (The folks at Edvotek recommend using a wooden coffee stirrer or a glass swizzle stick *[see Figure 25.1].*)

Gingerly twirl back and forth with the tip of the stick suspended just below the boundary between the alcohol and the buffer solution. Longer pieces of DNA will then spool onto the stick, leaving smaller molecules behind. After a minute of twirling, pull the stirrer up through the alcohol, which will make the DNA adhere to the end of the stick and appear as a transparent viscous sludge clinging to the tip.

Although these results are impressive, this simple and inexpensive procedure does not yield a pure product. Professionals add enzymes that tear apart the RNA molecules to make sure they do not get mixed up with the coveted DNA.

Figure 25.1 A kitchen laboratory includes most of the items needed to isolate DNA. A drinking straw, for example, can be used to add alcohol to the solution (a), and a coffee stirrer serves to spool the DNA (b).

Even after the most thorough extraction, some residual DNA typically lingers in the vessel, forming an invisible cobweb within the liquid. But with a little more effort, you can see that material, too. Some dyes, like methylene blue, will bind to charged DNA fragments. A tiny amount added to the remaining solution will stain tendrils of uncollected DNA. I don't know whether any household dyes, like food coloring or clothing or hair dyes, will also work, so I invite you to find out. Add only a drop: you want all the dye molecules to bind to the DNA, with none left over to stain the water.

Exciting as it may be, extracting an organism's DNA is only the first step in most biological experiments. You'll probably want to learn what further investigations you can do—for example, sorting the various DNA fragments according to their lengths. *[See Chapter 2 to see how to separate the fragments using electrophoresis. Ed.]*

The Society for Amateur Scientists and Edvotek have joined forces to create a kit containing cell samples, lab ware, enzymes, buffers, and detergents that can help you create higher-quality preparations. To order, send $35 to the Society for Amateur Scientists at 5600 Post Road, #114-341, East Greenwich, RI 02818.

GLOSSARY

abbe condenser   A common type of microscope stage condenser, usually consisting of a color-corrected lens doublet and an adjustable iris.

absolute temperature   Temperature is a measure of the average kinetic energy of a group of interacting molecules. If all of the kinetic energy could be removed from a substance (an ideal that can be approached but not actually achieved), that substance's temperature would be what we call "absolute zero." This ideal defines zero on the absolute temperature scale. An absolute temperature is any temperature measurement that is made relative to absolute zero.

aerotaxis   In botany, refers to a plant's sensitivity to gases.

agar   A medium, manufactured from algae, with the consistency of stiff Jell-o. When infused with the proper nutrients, bacteria and algae can grow on agar, and so it is often used to culture these organisms. Agar is also often used in electrophoresis.

agouti   A property of some types of fur. It denotes a characteristic variegated appearance, caused by the fact that each hair has a light and dark portion.

algae   Common types of single-celled plants.

algaecide   Any substance that kills **algae.**

amino acid   Any of 20 different chemicals that are the building blocks of proteins. All proteins of all life on earth are constructed entirely from these 20 building blocks.

amoeboid slime mold   A common term for fungi having no definitive shape, belonging to the order Myxomycetes.

angstrom   A unit of distance equal to one ten billionth (10^{-10}) of one meter. Atoms are approximately one angstrom wide.

angularity   A measure of the total of all angular bends throughout an object.

annular illumination   Illumination that is directed into the shape of a cone.

aseptic   Free from microscopic organisms including bacteria and viruses.

autoclave   A device used to sterilize instruments. It is a pressure chamber that allows one to raise the temperature of water above 100 degrees C without the water boiling. These higher-than-boiling temperatures readily kill microorganisms.

auxin   A botanical hormone that adjusts the rate of growth of plant cells. It stimulates the growth in green stem and leaf cells, and retards growth in root cells.

bacteriostatic drugs   Any substance that stops the growth or multiplication of bacteria.

Beer's law   The proposition that the light absorbed by a homogeneous liquid is linearly proportional to the thickness of the liquid. This is only true if the thickness is so small that only a small fraction of the light is absorbed. Do not rely on Beer's law if less than about 60 percent of the light makes it through the liquid.

binary fission   The process by which many microscope organisms reproduce by splitting in two.

biotechnology   Any technology that relies on the chemistry of life to achieve its ends.

botany   The study of plants.

bourdon gauge   A device used to measure low pressures.

bright-field lighting   In microscopy, the technique of illuminating the subject with a uniform bright light.

British thermal unit (BTU)   A measure of thermal energy equal to 1,055 joules.

buffer   A liquid formulated to keep the pH constant during the course of some chemical reaction taking place within it.

candle, standard international   A measurement standard used to quantify an object's brightness. It was originally defined as the energy flux produced by a ⅙-pound candle of sperm wax, burning at a rate of 120 grains per hour. This standard was replaced in 1921 by one based on incandescent lamps. Candles are no longer used as brightness standard.

capillitium   In botany, a network of hairlike filaments in which the sporules of some fungi are retained.

carbohydrate   Any of certain organic compounds composed of carbon, oxygen, and hydrogen, including the sugars, starches, and celluloses.

carbon dioxide   A chemical whose molecules are composed of one atom of carbon and two of oxygen. At standard temperature and pressure, carbon dioxide is a gas.

Carboniferous epoch   The great coal-making period in earth's history. It occurred in the Paleozoic era during which the warm, damp climate produced vast forests that later formed rich coal seams. It took place about 360 and 290 million years ago.

centrifuge A device used to separate materials suspended in a liquid by the use of centrifugal forces.

chemotaxis In botany, sensitivity of a plant to chemicals.

chromatic aberration In optics, a distortion of an image created when lights of different colors are refracted to different distances along the focal axis.

chromatography A method used to separate mixtures of molecules in a solution. Chromatography relies on the fact that different substances migrate through materials at different rates to carry out the separation.

chromosomes In living cells, the structures that contain the genes.

clearing In microscopy, to prevent decay a specimen of living tissue is usually treated with a substance, like alcohol, that is incompatible with life. (See **fixation**.) Afterward, the alcohol is "cleared" or removed by the addition of a solvent, like xylene, into which resins will dissolve that harden to create a durable specimen that can be viewed for decades.

clinostat A device that simulates weightlessness for the purpose of growing plants.

colorimeter A device used to characterize the chemical composition of a liquid by measuring the transmittance of colors through the liquid.

condenser See **microscope, condenser.**

dark field lighting In microscopy, the practice of illuminating an object from the side only so that it appears as brightly lit against a dark background.

Davis diaphragm An iris, a device that opens and closes to control the amount of light that passes through it.

deoxyribonucleic acid (DNA) The molecule of inheritance. DNA contains the genetic information used by cells to construct the proteins needed to carry out all of the cell's functions.

diatom A single-celled organism surrounded by an exoskeleton of silica dioxide.

Drosophila Scientific name for the common fruit fly.

electrophoresis Any technique that uses electric currents to separate molecules within a solution.

electrophoresis, gel-based Electrophoresis that occurs when the molecules to be separated are forced to move through gel.

electrophoresis, paper-based Electrophoresis that occurs when the molecules to be separated are forced to move through paper.

Erlenmeyer flask A triangularly shaped, round-bodied flask.

fixation   In microscopy, to prevent decay a specimen of living tissue is usually treated with a substance, like alcohol, that is incompatible with life.

flagellates   Any microscopic organism that propels itself with a flagellum, or whiplike tail.

flatworm   Any of a large group of worms with a flattened, unsegmented body. Many flatworms are parasitic.

fluorescent lamp   A widely used type of light. Fluorescent lamps are notable because they produce a bright light with relatively little heat.

fluorescent lamp, ballast   An electrical device that creates a current pulse used to initiate the plasma inside a fluorescent lamp.

footcandle   A measure of brightness. A footcandle is the energy flux cast by one standard international candle onto a surface one square foot that is one foot away.

fungi   A large group of microscopic plants having no green color. The group includes the mushrooms, toadstools, and microscopic plants that grow on other plants, such as molds, mildew, smut, rust, brand, and so on.

furnace   Any device that can reach temperatures higher than about 600 degrees F.

gamete   A reproductive cell that can combine with a complementary cell to form a new cell that develops into a new organism.

genetics   The study of genes.

geotropism   In biology, sensitivity of an organism to the force of gravity.

giant amoeba   An amoeba that is large enough to be seen with the naked eye. The species *Pelomyxa carolinesis* can grow up to 5 millimeters.

Gibberella fujikuroi (fungus)   The fungus from which **gibberellic acid** is derived.

gibberellic acid   A substance, originally isolated from a fungus, that dramatically stimulates plant growth.

greenhouse   An enclosure with controlled temperature and humidity used to raise plants.

grow light   A **fluorescent lamp** designed to approximate the solar spectrum more closely than an ordinary fluorescent lamp.

growth inhibitors   In botany, any substance that retards growth.

Hall effect transducer   A device that used the "Hall effect" to measure magnetic fields.

heterozygote   In Mendel's theory of heredity, a plant or an animal that has one or more recessive characteristics and hence is not breeding true to type; hybrid.

high pass filter   In electronics, a device that passes frequencies higher than its designed frequency, and blocks all signals with lower frequencies.

horticulture   The art or science of cultivating a garden.

hydra   Any of a group of freshwater microscopic animals with a tubelike body and a mouth surrounded by tentacles.

hydroponics   Collectively, all techniques used to grow plants in a fluid-based medium without the benefit of normal soil.

hydroponics, gravel culture   Use of gravel as a bed in a hydroponic garden.

hydroponics, sand culture   Use of sand as a bed in a hydroponic garden.

hydroponics, water culture   A water hydroponic culture that uses no solid material to anchor the roots.

incandescent lamp   Any lamp that generates light using a heated tungsten filament.

incubator   A temperature-controlled chamber used to grow bacteria, virus, or algae.

inhibition zone   An area where no new isolate will grow.

Koehler illumination   A mode of microscope illumination in which the light source is imaged onto the condenser iris diaphragm and the field of the diaphragm (in front of the lamp collector lens) is imaged by the condenser onto the plane of focus of the specimen. With Koehler illumination, the aperture and field can be regulated independently to provide maximum resolution and optimum contrast. Also, Koehler illumination generates a field with uniform illumination that is circumscribed by the image of the field diaphragm.

kymograph   A special device that records biological activity, like respiration or bioelectricity, on a graph.

Lepidoptera   Scientific name for the order containing all butterflies and skippers.

low pass filter   In electronics, a device that passes frequencies lower than its designed frequency, and block all signals with higher frequencies.

manometer   A device used to measure atmospheric pressure.

metabolic rate   A measure of the rate at which living organisms burn chemical energy.

metabolism   The process by which living organisms burn chemical energy. It is measured by observing the rate at which an organism consumes oxygen and releases carbon dioxide.

microbiology   The study of microscopic life.

microcentrifuge   A centrifuge designed to work with special small sample vials.

microscope   A device that magnifies tiny objects so they can be observed and studied.

microscope, condenser   In a transmission microscope, the lens system in the base that holds the specimen slide. These lenses concentrate and otherwise manipulate the light that illuminates the specimen.

microscope, electron   A microscope that uses a scanning electron beam to create an image of a microscopic subject.

microscope, phase contrast   A microscope that enhances the contrast of a microscopic specimen by converting phase differences in light coming from different parts of a specimen into amplitude differences in light. This works because the human eye is not sensitive to the phase of the light, but it is sensitive to the amplitude.

microscope, transmission   Any microscope that illuminates specimens by passing light through them from behind and into the objective lens.

microscope, video   Any microscope that can record a video image.

mold   A downy or furry growth on the surface of organic matter, caused by fungi, especially in the presence of dampness or decay.

molecular biology   The subdiscipline of biology that studies the molecular basis of life.

monolayer culture   A culture of animal tissue consisting of a single layer of cells.

mycetozoa   A group of protozoans, most of which are in the family of slime molds.

Mylar   A type of plastic noted for its tensile strength.

myxamoeba   A myxomycete in a certain stage of development.

myxomycete   Any of a class of slime molds, consisting of masses of naked protoplasm and having some of the characteristics of both plants and animals, but generally classified as plants (fungi).

objective lens   In a microscope or telescope or other optical system, the lens closest to the object being observed.

oblique bright-field illumination   In microscopy, the technique illuminating the subject with a bright light directed from a steep angle.

operational amplifier (op amp)   An extremely important kind of amplifier of electric signals in which the gain is set by negative feedback, that is, by adding part of the output back in to the amplifier's negative input.

paper chromatography   Chromatography in which paper is used as the separating medium. (See **chromatography.**)

paramecium   Any of a number of related single-celled, elongated animals having a large mouth in a fold in the side and moving by means of cilia.

patching A process used to establish new cell cultures that are identical copies of the original culture. A small square, typically about one millimeter on a side, is transferred to a freshly prepared culture medium and the incubation is continued.

Penicillium notatum The scientific name of the mold from which penicillin is derived.

peridium The outer bark and the layer of soft, growing tissue between the bark and the wood in plants.

Pfeiffer, Wilhelm A German plant physiologist.

pH A symbol for the degree of acidity or alkalinity of a solution. Originally, and often still, expressed as the logarithm of the reciprocal of the hydrogen ion concentration in gram equivalents per liter of solution (today it sometimes has other operational definitions); pH 7 (0.0000001 gram atom of hydrogen ion per liter) is the value for pure distilled water and is regarded as neutral; pH values from 0 to 7 indicate acidity, and from 7 to 14 indicate alkalinity.

phase In optics, the phase denotes a particular position inside a wave.

phase grating A grooved transparent material designed to create differences in the phase shift across its surface that it produces.

phase image An image of an object created by converting the phase shifts (which the human eye cannot see) that take place as light passes through a subject into amplitude shifts (which the eye can see).

phase plate A transparent plate that uniformly shifts the phase of the incident light by a constant amount.

phase-contrast microscopy See **microscope, phase contrast.**

photoperiod When experimenting with plants, one often needs to control the cycle of light and dark to which they are exposed. The photoperiod is the time for one such cycle.

photoperiodism Refers to the response of plants and animals to variations in the relative duration of day and night.

photosynthesis The process by which plants use sunlight to build the complex molecules necessary to sustain their lives.

phototaxis Sensitivity to light. Biologists say that an organism is phototaxic only if the entire organism responds to light. An organism that moves toward light is often said to be positively phototaxic, and negatively phototaxic if it moves away.

phototropism Refers to plants that move, bend, or grow toward a light source.

plant press A device used to press plants flat while drying them. It is used to convert a living fluid-filled plant into a dry specimen that can be stored in a museum.

plasma clot   A form of cultivated tissue specimen; especially good for use in microscopic examination.

plasmodium   A mass of protoplasm with many nuclei, formed by the fusion of many single-celled organisms. Or, any of various unicellular parasites found in red blood corpuscles, one variety of which causes malaria.

pond scum   Any alga that grows abundantly to form mats in ponds and streambeds.

proportional relief valve   Any valve that allows gas to flow through it at a rate that is proportional to the pressure difference across it.

protein   Any number of molecules that run virtually all of the biological processes of living cells. Proteins are composed of long chains of amino acids. Their recipes are recorded in a cell's DNA in areas called genes.

protozoa   A phylum of single-celled, usually microscopic, animals belonging to the lowest division in the animal kingdom. Protozoa usually reproduce by binary fission. The organs of locomotion are varied. In some of the more advanced forms, movements are effected by cilia, in others by flagella. But most push themselves along by extruding and retracting sections of protoplasm. Most protozoa thrive in water.

psychrometer   A device to measure relative humidity.

Ramsden disk   A small circular patch of light that appears at the eyepiece above the ocular lens. This is the exit pupil of an optical instrument that, in a microscope adjusted for **Koehler illumination,** lies in a plane conjugate with the objective rear focal plane, condenser iris, and light source. Alteration of the Ramsden disk (e.g., by the observer's iris) modifies the aperture function, diffraction pattern, and direction of view of the specimen.

rare-earth magnet   An especially powerful ceramic magnet that contains one of the 17 so-called rare-earth elements (scandium, yttrium, and 15 elements from lanthanum to lutetium).

relative humidity   A measure of the water content in air. Relative humidity is the concentration of water vapor in the air divided by the total amount of water vapor that the air at that pressure and temperature could support. It is usually reported as a percentage.

resolving power   A measure of the ability of an optical system to resolve two objects that are near each other. It is the smallest angle the objects can be separated by and still be clearly distinguished.

respiratory quotient   A measure of **metabolism.** The respiratory quotient is the ratio of the number of oxygen molecules consumed by an organism over a given time, divided by the number of carbon dioxide molecules released.

Rheinberg color illumination   A form of optical staining, Rheinberg illumination is a variation of **dark field lighting** that uses colored filters to provide rich colors to both the specimen and the background.

ribonucleic acid (RNA)   A chemical similar to a single strand of DNA. RNA delivers DNA's genetic message to the cytoplasm of a cell where proteins are made.

saline   Sterile salt water. Often, the salt concentration is made to match that of human blood.

sclerotium   In certain fungi, a hardened, weblike, black, or reddish-brown mass of threads in which food material is stored. Also, the state of the plasmodium when it has matured.

sodium hydroxide   A caustic substance whose chemical formula is NaOH.

spectra, first and second order   When a spectrum is generated by wave interference through a diffraction grating, the first order spectra are the first complete set of colors, the second order spectra are the complete second set, and so on.

spectral intensity   The brightness of a given part of a **spectrum.**

spectrum   The array of colors obtained by passing light through a prism or through a diffraction grating.

spirometer   An instrument that responds to the volume of oxygen consumed by a subject.

sporangia   The fruiting bodies that are part of the life cycle of some amoebic slime molds.

sporule   A small spore.

standard pressure   One atmosphere or 760 torr.

standard temperature   Zero degrees Centigrade or 273.15 Kelvin.

sterilize   To kill all living things on an object.

stomata   Pores in a plant that enable gas exchange to take place and water vapor to be lost.

sugar   A sugar is any of a number of carbohydrates with the fundamental chemical formula $C_{12}H_{24}O_{12}$. Three common sugars are glucose or "blood sugar"; galactose, which is found in milk; and fructose, which is found in honey. These so-called monosaccharides can be linked together to form larger molecules. The disaccharides, like sucrose (ordinary table sugar) and lactose (the sugar in milk), are formed by linking glucose and fructose, and glucose and galactose respectively. The polysaccharides are formed by linking many of monosaccharides into chains that can contain many thousands of atoms. Starches and cellulose are both examples of such chains.

temperature controller   Any device that provides feedback necessary to maintain something at a desired temperature.

temperature scale, Celsius (C)   The Celsius scale is defined by the triple point of water, and boiling point of water at standard pressure. The triple point of water is the temperature at which its three phases (solid, liquid, and gas) can all exist in equilibrium. It is defined to be exactly 0.01 degrees C. The boiling point of water at standard pressure is to occur at exactly 100 degrees C. With this definition, at standard pressure water freezes at very nearly 0 degrees C.

temperature scale, Fahrenheit   The Fahrenheit temperature is defined from the Celsius scale. The temperature in degrees Fahrenheit is 1.8 times the temperature in Celsius +32. With this definition, the freezing point of pure water at standard pressure is calculated to be approximately 32 degrees, and the boiling point is 212 degrees.

temperature scale, Kelvin   In the Kelvin, the divisions in temperature are exactly equal to the divisions in the Celsius scale; however, the scale sets its zero point at absolute zero temperature. On this scale, at standard pressure water freezes at approximately 273 Kelvin and boils at 373 Kelvin. Note: Although people often say "degrees Kelvin" in reciting a temperature, the accepted standard is to drop the word "degrees" as above.

terrarium   A glass or plastic enclosure in which small plants and animals are raised.

thermoperiod   In botany, the temperature of a greenhouse is often controlled so as to mimic the cycle of high and low temperature that takes place during a day. The time between successive cycles, called the thermoperiod, can be controlled by the experimenter and is sometimes set to something other than 24 hours.

torr   A pressure equal to that generated by a column of mercury 1 millimeter high.

transducer   Any device that converts one form of energy into another. A microphone, for instance, transduces sound energy into electrical energy.

tropotaxis   In botany, sensitivity of a plant to gravity.

turbidity   A measure of the cloudiness of fluid. It is typically measured by seeing how far light can penetrate the fluid.

UVB (ultraviolet B)   A band of ultraviolet radiation produced by the sun that is particularly effective at damaging DNA. Wavelengths range from 280 to 320 nanometers.

vacuum pump, dry-vane   A mechanical vacuum pump that exposes no lubricating oil to the vacuum. These pumps are not limited to the vapor pressure of lubricating oils, and they do not risk contaminating whatever is under vacuum with oil vapors.

vitamin Any of a number of unrelated complex organic compounds that are found in various foods and are essential in small amounts to maintain good health.

volt A unit of electrostatic potential. One volt is the potential required to drive one ampere of current through a resistance of one ohm.

Walburg apparatus A device used to measure **metabolic rate.**

watt In metric units, the measure of power. Something is said to generate one watt if it delivers one joule of energy every second.

wave interference The process by which two waves add together to either reinforce or cancel each other.

well slide In microscopy, a slide with a small depression used to hold the specimen.

zygote Any cell formed by the union of two **gametes.**

FURTHER READING

Janice VanCleave's A+ Projects in Biology: Winning Experiments for Science Fairs and Extra Credit, Janice Pratt VanCleave, John Wiley & Sons, 1993, ISBN: 0-471-58628-5

American Horticultural Society A to Z Encyclopedia of Garden Plants, Christopher Brickell and Judith Zuk, DK Publishing, 1997, ISBN: 0-7894-1943-2

Biography of a Germ, Arno Karlen, Schocken Books, 2000, ISBN: 0-375-40199-7

Botany for Gardeners: An Introduction and Guide, Brian Capon, Timber Press, 1992, ISBN: 0-88192-258-7

Fearsome Fauna: A Field Guide to the Creatures That Live in You, Roger M. Knutson, W.H. Freeman and Company, 1999, ISBN: 0-7167-3386-2

A Field Guide to Germs, Wayne Biddle, Anchor, 1996, ISBN: 0-385-48426-7

Stokes Guide to Observing Insect Lives, Donald Stokes, Little, Brown and Company, 1984, ISBN: 0316817279

Instant Notes in Microbiology, J. Nicklin, T. Paget, K. Graeme-Cook, R. A. Killington, J. Nickler, Springer Verlag, 1999, ISBN: 0-387-91559-1

Journey to the Ants: A Story of Scientific Exploration, Bert Holldobler and Edward Osborne Wilson, 1995, Belknap Press; ISBN: 0-674-48526-2

Methods in Plant Biochemistry and Molecular Biology, William V. Dashek, 1997, CRC Press, ISBN: 0-8493-9480-5

National Audubon Society Field Guide to North American Butterflies, Robert Michael Pyle, Knopf, 1996, ISBN: 0-394-51914-0

National Audubon Society Field Guide to North American Insects and Spiders, Lorus J. Milne and Susan Rayfield, Knopf, 1980, ISBN: 0-394-50763-0

Oxford Dictionary of Biochemistry and Molecular Biology, A. D. Smith, Oxford University Press, 2000, ISBN: 0-19-850673-2

Peterson First Guide to Caterpillars in North America, Amy Bartlett Wright and Roger Tory Peterson, 1998, Houghton Mifflin Company (paperback), ISBN: 0-395-91184-2

Plant Biochemistry and Molecular Biology, 2nd Edition, Peter J. Lea and Richard C. Leegood, John Wiley & Sons; 1999, ISBN: 0-471-97683-0

Plant Biology Science Projects, David R. Hershey, John Wiley & Sons, 1995, ISBN: 0-471-04983-2

Plant Microtechnique and Microscopy, Steven E. Ruzin, Oxford University Press; 1999, ISBN: 0-19-508956-1

The Amateur Naturalist, Gerald Malcolm with Lee Durrell, Knopf, 1983, ISBN: 0-394-53390-9 (Out of print, but wonderful!)

Scientific American's "The Amateur Scientist"—the Complete 20th Century Collection on CD-ROM, Shawn Carlson and Sheldon Greaves, Tinkers Guild, 2000, ISBN: 0-970-34760-X

The Practical Entomologist, Rick Imes, Fireside, 1992, ISBN: 0-671-74695-2

SUPPLIER LIST

Berkshire Biological Supply, 800-462-1382, *www.berkshirebio.com*

BioQuip, 17803 LaSalle Ave., Gardena, CA 90248, 310-324-0620, *www.bioquip.com*. General biological supplies

BioSupplyNet, Web: *biosupplynet.com*

Blue Spruce Biological Supply, 701 Park Street, Castle Rock, CO 80104, 800-825-8522, *www.bluebio.com*

Carolina Biological Supply Co., 2700 York Road Burlington, NC 27215, 800-334-5551, *www.carolina.com*

Connecticut Valley Biological Supply Co., 82 Valley Road, Southampton, MA 01073, 413-527-4030 or 800-628-7748, *www.ctvalleybio.com*

Edmund Scientific Co., 60 Pearce Avenue, Tonawanda, NY 14150-6711, 800-728-6999, *www.edsci.com*

Fisher Scientific, 4500 Turnberry Drive, Hanover Park, IL 60103, 800-766-7000, *www.fishersci.com*

Frey Scientific Company, 100 Paragon Parkway, Mansfield, OH 44905, 800-225-3739, *www.freyscientific.com*

Pacific Biological Supply, 15035 Ventura Boulevard, Sherman Oaks, CA 91403, 877-588-0310, *www.pacificbio.com*

Society for Amateur Scientists, 5600 Post Road, Suite 114-341, East Greenwich, RI 02818, 401-823-7800 *www.sas.org*

Trans-Mississippi Biological Supply, 590 Cardigan Road, St. Paul, MN 55126, 800-544-5901, *www.tmbs.com*

WARD's Natural Science Establishment Inc., 5100 West Henrietta Road, Box 92912, Rochester, NY 14692, 800-962-2660, *www.wardsci.com*

Worldwide Butterflies Ltd., Compton House Nr., Sherbourne Dorset DT9 4QN, UK, +44 (0) 1935 474608, *www.wwb.co.uk*

ORGANIZATIONS

The Amateur Entomological Society, AES, P.O. Box 8774, London SW7 5ZG, UK, *www.theaes.org*

Amateur Entomologist, *http://pages.infinit.net/laurentl/index.html*

American Society of Plant Physiologists, 15501 Monona Drive, Rockville MD 20855-2768, 301-251-0560, *www.aspb.org*

Butterfly World, Tradewinds Park, 3600 West Sample Road, Coconut Creek, FL 33073, 954-977-4400, *www.butterflyworld.com*

Calloway Gardens Day Butterfly Center, U.S. Highway 27, Pine Mountain, GA 31822, 800-282-8181, *www.callawaygardens.com*

Entomological Society of America, 9301 Annapolis Road, Lanham, MD 20706-3115, 301-731-4535, *www.entsoc.org*

Lepidoptera Research Foundation, 9620 Heather Road, Beverly Hills, CA 90210-1757, *www.doylegroup.harvard.edu/~carlo/JRL/subscribe.html*

Lepidopterists' Society, Ernest H. Williams, Department of Biology, Hamilton College, Clinton, NY 13323, *www.furman.edu/~snyder/snyder/lep*

McCrone Research Institute, 2820 South Michigan Avenue, Chicago IL 60616-3292, 312-842-7100, *www.mcri.org* (the world's leading teachers of microscopy techniques)

Microscopy Society of America, Bostrom Corp., 230 East Ohio, Suite 400, Chicago, IL 60611, 800-538-3672, *www.msa.microscopy.com*

The Royal Entomological Society of London, 41 Queen's Gate, London, SW7 5HU, UK, 020-7584-8361, *www.royensoc.demon.co.uk*

Society for Amateur Scientists, 5600 Post Road, Suite 114-341, East Greenwich, RI 02818, 401-823-7800, *www.sas.org*

Xerces Society, Department of Zoology and Physiology, University of Wyoming, Laramie, WY 82071, *www.xerces.org*

INDEX

environment controls, 90
experiment needs, 88–90
gibberellic acid, 87–88
metabolism measurement, 91–93
oxygen consumption, 93–95
recordkeeping, 90–91
ultrasonics, 96–102 (*see also*
ultrasonics)

H
Hadorn, Ernst, 184
hairless mouse, 182–183
Haldane, J. S., 22
Hall, Francis C., 54
Hall effect transducer, 166, 167–169
hanging-drop preparation, slime
molds, 123
Hansen, Stephen P., 26–27, 28
Hayward, Roger, 135
heartbeat, insects, 165–169
heaters, high altitude chamber, 30
heating cables, greenhouses, 61
HeLa cell line, 111, 116
herbicides, 81
high altitude chamber, 25–30
Honeywell Micro Switch, 167
Honeywell transducer, 157
Horowitz, Evalyn, 96–97
horseradish, 82
humidity
greenhouses, 57–59, 61
slime molds, 123–124
hybrid corn, 181
Hydra viridissima, phototaxis, 138
hydroponics, 62–69
forms of, 64–65
history of, 62–63
nutrients, 63–64, 66–69
water, 65–67

I
import restrictions, lepidoptera,
143–144
incandescent light
greenhouse, 55–56, 57–58, 60, 61
growth stimulants (botany), 90
ultrasonics and plant growth, 97,
102

incubation
animal tissue cultures, 113–114,
115, 118
slime molds, 128
Indian moon moth, 142
3-indoleacetic acid, 100–102
inertia, geotropism, 74, 76
Ingalls, Albert G., vii, x
inorganic salts, animal tissue cultures,
112
insecticides, 82
insect metabolism measurement,
153–159. *See also* metabolism
measurement
equipment, 153–156
experiments, 157–159
insect preservation, 160–164. *See also*
drying; preservation
instrumentation amplifiers, 167
insulin, genetics, 180
intelligence, of animals, 170
iron, hydroponics, 63
isopropyl alcohol, 8, 162, 163, 198
isoxanthopterin, 193

K
killing jar, 145, 149, 151
kymograph
metabolism measurement, 20
tin-can, 31–35

L
labeling, lepidoptera, 145–147
LaFond, Richard, 184
Lauber, Jean K., 12
laundry detergent, 197
Lawrence, Robert, 87
leaching, sea plant preservation, 53
lemon peel, 82
lepidoptera, 141–152
breeding of, 143–144, 149
classification, 148
collection techniques, 144–145,
148–149
mounting and labeling, 145–147,
149, 151
silkworms, 141–142
lethal mutation, genetics, 179–180